Test-Driving the Future

PHILOSOPHY, TECHNOLOGY AND SOCIETY

SERIES EDITOR: SVEN OVE HANSSON

Technological change has deep and often unexpected impacts on our societies. Sometimes new technologies liberate us and improve our quality of life, sometimes they bring severe social and environmental problems, sometimes they do both. This book series reflects philosophically on what new and emerging technologies do to our lives and how we can use them more wisely. It provides new insights on how technology continuously changes the basic conditions of human existence: relationships among ourselves, our relations to nature, the knowledge we can obtain, our thought patterns, our ethical difficulties, and our views of the world.

Titles in the Series

The Ethics of Technology: Methods and Approaches, edited by Sven Ove Hansson

Nanotechnology: Regulation and Public Discourse, edited by Iris Eisenberger, Angela Kallhoff, and Claudia Schwarz-Plaschg

Water Ethics: An Introduction, by Neelke Doorn

Humans and Robots: Ethics, Agency, and Anthropomorphism, by Sven Nyholm

Interpreting Technology: Ricoeur on Questions Concerning Ethics and Philosophy of Technology, edited by Mark Coeckelbergh, Alberto Romele, and Wessel Reijers

The Morality of Urban Mobility: Technology and Philosophy of the City, by Shane Epting

Problem-Solving Technologies: A User-Friendly Philosophy, by Sadjad Soltanzadeh

Test-Driving the Future: Autonomous Vehicles and the Ethics of Technological Change, edited by Diane Michelfelder

Test-Driving the Future

Autonomous Vehicles and the Ethics of Technological Change

Edited by Diane Michelfelder

ROWMAN & LITTLEFIELD
Lanham • Boulder • New York • London

Published by Rowman & Littlefield
An imprint of The Rowman & Littlefield Publishing Group, Inc.
4501 Forbes Boulevard, Suite 200, Lanham, Maryland 20706
www.rowman.com

86-90 Paul Street, London EC2A 4NE

Copyright © 2022 by Diane Michelfelder

All rights reserved. No part of this book may be reproduced in any form or by any electronic or mechanical means, including information storage and retrieval systems, without written permission from the publisher, except by a reviewer who may quote passages in a review.

British Library Cataloguing in Publication Information available

Library of Congress Cataloging-in-Publication Data
Names: Michelfelder, Diane P., 1953- editor.
Title: Test-driving the future : autonomous vehicles and the ethics of technological change / edited by Diane Michelfelder.
Description: Lanham : Rowman & Littlefield, [2023] | Series: Philosophy, technology, and society | Includes bibliographical references and index.
Identifiers: LCCN 2022030738 (print) | LCCN 2022030739 (ebook) | ISBN 9781786613233 (hardcover) | ISBN 9781786613240 (epub) | ISBN 9781538173619 (paperback)
Subjects: LCSH: Automated vehicles—Moral and ethical aspects. | Automated vehicles—Forecasting.
Classification: LCC TL152.8 .T47 2023 (print) | LCC TL152.8 (ebook) | DDC 629.2—dc23/eng/20220822
LC record available at https://lccn.loc.gov/2022030738
LC ebook record available at https://lccn.loc.gov/2022030739

Contents

Preface vii

Acknowledgments ix

Chapter 1: Test-Driving the Future: Ordinary Environments and the
Ethics of Technological Change 1
Diane Michelfelder

Chapter 2: A Postphenomenology of Electric, Self-Driving, and
Shared Vehicles 15
Galit Wellner

Chapter 3: The Ethics of Crossing the Street 29
Robert Kirkman

Chapter 4: Who Is Responsible If the Car Itself Is Driving? 43
Sven Ove Hansson

Chapter 5: The Ethics of Transitioning Toward a Driverless Future:
Traffic Risks and the Choice Among Cars with Different Levels
of Automation 59
Sven Nyholm

Chapter 6: Stop Saying That Autonomous Cars Will Eliminate
Driver Distraction 77
Robert Rosenberger

Chapter 7: Automated Vehicles and Environmental Justice:
Addressing the Challenges Ahead 97
Shane Epting

Chapter 8: Vehicles of Change: The Role of Public Sector
Regulation on the Road to Autonomous Vehicles 109
Patrick Schmidt and Jeremy Carp

Chapter 9: Planes, Trains, and Flying Taxis: Ethics and the Lure of
    Autonomous Vehicles   131
    Joseph Herkert, Jason Borenstein, and Keith W. Miller

Chapter 10: Experiencing the Future: A Phenomenological
    Exploration of Automated Vehicles   151
    Ike Kamphof and Tsjalling Swierstra

Index   167

About the Contributors   173

Preface

We can trace the wish for a future in which autonomous vehicles are the norm back to another wish—the wish to be free from vehicular accidents while traveling on the roadways. In the context of this hope, vehicle safety powerfully emerges as the foremost ethical value to cultivate as this means of transportation develops and rolls out. From this perspective, the meaning of "test-driving the future" is centered first and foremost on autonomous vehicles as technological devices connected to and supported by other technologies that themselves are grounded in safety as a high-priority ethical value. Among these technologies are communicative systems, electrical systems, and of course the transportation infrastructure itself.

There is, however, more than one way to look at the meaning of the expression "test-driving the future" as it relates to autonomous vehicles. We can also consider autonomous vehicles to be technological imaginaries: imaginaries that are entangled in symbolic forms such as narrative, storytelling, culturally specific communicative conventions (rather than communicative systems), and familiar, everyday acts, practices, and environments involving autonomous vehicles, passengers, pedestrians, and others. If we explore these entanglements and how we are also taken up by them, then a different set of ethical values and issues, as well as different design spaces, could emerge that are clouded over when the starting point for "test-driving the future" is taken to be the autonomous vehicle itself.

The main purpose of this collection of essays is to explore what it might be to "test-drive the future" in this second sense. But there is another purpose to this book as well: to be a prism for getting an understanding of what ethically related questions might be important to ask in the context of technological change in general, particularly technological change whose possibilities for transforming significant aspects of individual and societal life loom large. What does "test-driving the future" in the second sense have to say about this?

Acknowledgments

I want to thank all of those involved in the making of this book, starting with Sven Ove Hansson, the editor-in-chief of the Philosophy, Technology and Society book series in which this volume appears, who was unswervingly enthusiastic and supportive from our initial discussions through the conclusion of this project.

I also would like to recognize the editors at Rowman & Littlefield International with whom I had the pleasure of working at various times. My appreciation goes out to Isobel Cowper-Coles for her interest in this volume and for soliciting three anonymous referees whose comments on the original proposal helped to strengthen this work. My gratitude also goes out to Scarlet Furness and Frankie Mace. Because of the COVID-19 pandemic, this volume took longer to complete than originally planned. Scarlet and Frankie met my requests for deadline extensions with most welcome grace and kindness. A big thank you as well to Sylvia Landis and Janice Braunstein for their abundant thoughtfulness and expertise in guiding this volume into and along the production process.

Finally, my gratitude goes out to each of the contributors to this volume for their show of spirit and understanding in sticking with this project during unexpectedly difficult times and for their abundant and fruitful insights into the ethics of autonomous vehicles. I hope they will lead those reading this book to more of the same.

The delay caused by the COVID-19 pandemic gave this book more time to simmer, as it were, letting its chapters work together to produce a different and more unified whole than was anticipated in its early stages. While I am grateful for this outcome, it certainly was not worth the cost.

Chapter 1

Test-Driving the Future

Ordinary Environments and the Ethics of Technological Change

Diane Michelfelder

The essays assembled in this volume consider some underexplored ethical issues associated with the emergence of an innovation that regularly goes by more than one name. Most frequently, as in the title of this book, this innovation is called an "autonomous vehicle." Sometimes it is designated as a "driverless vehicle." On other occasions it is referred to as a "self-driving vehicle." While there are other names as well, these are the ones that are most in play.[1] In academic discourse and popular media alike, these expressions are often used as though one could be substituted for the other with no loss of meaning (see, for example, Lin 2016). The differences among these ways of talking tend to get glossed over as not mattering very much. Still, each one highlights a distinctive perspective worthy of philosophical consideration.

To speak about a driverless vehicle is to emphasize the place of this innovation within a lineage and trajectory of personal mobility systems. The historical narrative offered by such a framework hearkens back to the advent of the horseless carriage. At the same time, it gestures toward a future in which such vehicles appear as the natural extension of the latter through having eliminated the need to directly depend on living creatures in the shape of carbon-based and mistake-prone human beings guiding the vehicle by steering a wheel.

While the phrase "self-driving vehicle" also situates cars, trucks, buses, and other types of vehicles within the context of transportation and mobility systems, it does not do so from a perspective that involves a historical narrative but rather one that emphasizes automaticity. The expression calls attention to how these vehicles could be seen as artificial agents, capable of gathering data, processing it, and acting upon it, even though they are mindless and, as a result, their actions are not driven by intentionality. Still, as Sven Nyholm and Jilles Smids have pointed out (2020), theirs is an artificiality that is constantly updating itself in light of new information about their surroundings.[2]

To talk about "autonomous vehicles" is to bestow a special focus on the decisions that such cars and other means of transportation need to make while on the roadways. Put another way, the term identifies the driving scenario as one that is organized around the algorithmic-based judgments that these vehicles, as artificial products made possible through technical skill, to borrow the language of Yuk Hui (Hui 2016, 8), are constantly making while in use. This term at the same time also gestures toward the fact that their decisions are embedded and entangled in a wide array of systems and institutions on which they depend and which they also influence. Among others, these systems include political/legal/regulatory systems, workforce and labor systems, economic systems, and energy systems, as well as digital communication and data storage systems.

Decisions that promote the ethical value of human safety stand front and center when it comes to thinking about the decisions that autonomous vehicles need to make. Anyone who is even the least familiar with autonomous vehicles knows that the justificatory stress on safety—the "safety argument," as Daniel J. Hicks (2018) has called it—hinges on the desire to lessen the risks of driving by reducing the numbers of traffic fatalities and other accidents. At least within the United States, these numbers are showing little signs of going down. A look at data from the National Highway Traffic Safety Administration reveals that from 2019 to 2020 traffic fatalities rose by 7.2 percent to 36,680, their highest level since 2007 (NHTSA 2021), even while Americans were driving less due to the pandemic; and in the first nine months of 2022 the percentage of fatalities saw the greatest increase since this statistic began to be recorded (NHTSA 2022).[3]

Seen from this perspective, the meaning of "test-driving the future" is primarily linked to autonomous vehicles themselves, understood as technological artifacts that are connected to and supported by other technologies that also recognize safety as a high-priority ethical value. These include the electrical and communicative systems mentioned previously, as well as cybersecurity systems, the computational systems by which the algorithms used by autonomous vehicles are designed, and of course traffic signals, directional signs, and other elements of a transportation system itself.

In a pioneering as well as provocative essay, Patrick Lin (2016) argued that the development of autonomous vehicles needed to be ethically informed. As evidence, he pointed to reasons as to why autonomous vehicles would still be involved in crashes due to slippery road conditions and the like and might have to make difficult decisions as to what course of action to take (go straight and hit X or swerve and hit Y) if an accident could not be avoided. The possibility of such "trolley problem scenarios" would force engineers to include in the design of autonomous vehicles "crash-optimization algorithms" that could for instance prioritize passenger safety over the safety of pedestrians or bicyclists or prioritize running into to a more "crashworthy" automobile like a Volvo over striking another make of car known to be less safe—decisions that cannot be made without reference to human values.

Lin's thoughts have in their wake opened up extensive and lively debates. One set of debates accepts Lin's trolley problem framework and pursues the question of just what values ought to be incorporated into the algorithms that an autonomous vehicle could use in risky driving situations (see, for example, Lawlor 2022; Davnall 2020; Leben 2017); within this context, the question of what values would be acceptable to the public at large also comes up (Bonnefon et al. 2016). Taking a step back from this framework, another set of debates questions to what extent, if at all, trolley problem scenarios can lead to concrete directions for determining what autonomous vehicles ought to decide to do in situations where crashes are unavoidable (Lundgren 2021). In both cases though, safety is taken to be top priority. And once we begin to map the ethical issues associated with autonomous vehicles as seen through the lens of the value of safety, along the lines of what Mittelstadt et al. (2016) have done for the ethics of algorithms in general, we would discover a rich terrain of "spin-off" safety-related issues and values stretching well beyond determining a socially satisfactory solution to trolley problems in terms of accident algorithms.

Standing out as the closest spin-off would be the value of security, as this value is tightly connected to the importance of keeping vehicles safe from cyberterrorist attacks by malicious actors intent on causing harm and disruption in a variety of ways to a variety of objects, including individual vehicles, groups of vehicles networked together by communication systems, and transportation systems in general. Autonomy and privacy would count among the less tightly connected but still important spin-off values. In their comprehensive overview of ethical issues posed by autonomous vehicles, Hansson, Belin, and Lundgren (2021) observe that should autonomous vehicles prove to be highly effective in reducing accidents, it could potentially become illegal to drive a conventional car, a measure that would negatively impact a motorist's ability to drive a car of their choice. Even privacy could be seen as an element in the terrain of spin-off safety-related issues, as users of autonomous vehicles might need to give up some of their locational privacy for the sake of keeping others safer on the road.

In sum, looking at an autonomous vehicle as a technological artifact embedded within other technological systems puts the artifact itself on center stage, a space that, as Albrecht Fritzsche (2021) has suggested, "exposes" it and so provides a point of orientation for thinking about its place in society. As just noted, that point of orientation is the perspective of the safety argument. But, as Fritzsche spells out, once an artifact is exposed in this way, the possibility opens up to realize what is missing from how it is staged.

Just what is missing from thinking about the ethics of self-driving technologies from prioritizing the safety argument, and how might it open up another way of looking at the meaning of "test-driving the future"? As its title suggests, in "Enculturating Algorithms" (2019, 135), Raphael Capurro calls attention to how algorithms in general and mobility systems in particular are embedded within cultural norms and informal practices that vary from society to society. Such a perspective offers a clue to how we could take a starting point other than artifactuality for looking at the ethics of autonomous vehicles. We can look at them as technological imaginaries: imaginaries that are entangled in symbolic forms such as narrative, storytelling, culturally specific communicative conventions (in addition to communicative systems), and familiar, everyday acts, practices, and environments involving autonomous vehicles and their passengers, pedestrians, and others. Taking up such a starting point would not be to discount the value of safety, which is of course critical. But if we explore these entanglements and how we are also taken up by them, other ethical values and issues can emerge that are clouded over when the starting point for "test-driving the future" is taken to be the artifact of the autonomous vehicle itself. Shifting the focus like this turns the emphasis of ethical analysis away from the abstract and toward the details and contestations of ordinary environments themselves (see also Himmelreich 2018).[4]

CHAPTER PREVIEWS

In chapter 2, "A Postphenomenology of Electric, Self-Driving, and Shared Vehicles," Galit Wellner looks at the possibilities for autonomous vehicles by situating their development within the narrative history of horse-drawn carriages, but with a twist. Rather than focusing on that story as a linear progression, she invites us to consider three "technological families" connected to autonomous vehicles: electric engines, self-driving mechanisms, and shared ride algorithms. The lens of "technological families" brings out the "entanglements" among humans, technologies, and their surroundings. It shows how the "technological intentionality" (Peter-Paul Verbeek) afoot in the design of autonomous vehicles not only mediates human-world relations but also trumps our own intentionality in making decisions for us. Additionally, it points to a need for regulatory emphasis on autonomous vehicles as shared ride vehicles so that by means of this technology the world can become a better place.

Robert Kirkman follows in chapter 3 with "The Ethics of Crossing the Street." Crossing the street is such a basic human action that when we do it we tend not to think we are doing anything that has an ethical dimension to it. Turning to phenomenology along with Nel Nodding's ethics of care, Kirkman shows us that it does. Phenomenology offers the insight that when someone with the right of way wants to cross a street in front of a driver turning right, both actors need to communicate their intent in a negotiation of mutual recognition. The ethics of care shows that this negotiation is rooted in the ethical value of vulnerability. Playing out the same scenario with an autonomous vehicle shows that no shared space would be created, as the vehicle lacks the capability to recognize a pedestrian as a person. This leads Kirkman to worry that from a social constructivist point of view the introduction of autonomous vehicles into public spaces could reorganize them to be hostile, not life-affirming, spaces for those crossing the street.

Working through a series of illustrations in chapter 4, "Who Is Responsible If the Car Itself Is Driving?," Sven Ove Hansson focuses on answering this question through the lens of "causal singularization." When we pick out "the" cause from several distinct causal factors, and we believe that someone is blameworthy for causing something bad to happen because we do not approve of the values on which their action is based, we engage in the process of causal singularization. Put very generally, causal singularization is an ethical phenomenon. It lets us connect the dots among different kinds of dangerous actions, who should be held responsible for them, and in what way. Hansson identifies a social trend here as reflected by the Vision Zero project, which aims to increase traffic safety by placing blame responsibility for accidents on companies rather than consumers. This trend gives reason to expect that just as driving is "outsourced" in driverless vehicles, responsibilities for traffic safety will follow suit.

Sven Nyholm implicitly picks up on this point in chapter 5, "The Ethics of Transitioning Toward a Driverless Future: Traffic Risks and the Choice Among Cars with Different Levels of Automation," by posing two options as this transition gets underway. One is that the roads in such a future will include driverless vehicles operating at all six levels of autonomy—from no autonomy to full autonomy—as described by the Society of Automotive Engineers. The other is the position taken by Sparrow and Howard (2017), along with Elon Musk, that once fully autonomous vehicles become safer than all the other vehicles on the road, the other vehicles should be barred from the highways. Drawing on a variety of sources, including Sven Ove Hansson's defense of the riskiness of conventional driving, Nyholm defends the first of these options. He suggests, however, that allowing a mixture of vehicles on the road is justifiable only if the additional risks are "offset" through ongoing safety improvements to conventional vehicles.

With his contribution, "Stop Saying That Autonomous Vehicles Will Eliminate Driver Distraction" (chapter 6), Robert Rosenberger likewise strikes a cautionary note here with respect to decelerating the pace at which regulations are designed and implemented to eliminate driver distraction due to mobile devices. Such thinking, in Rosenberger's mind, represents an example of what he calls "spectatorial utopianism." The mindset of "spectatorial utopianism," he argues, is fueled by holding on to an abstract concept of technology plus a belief in the narrative of technological progress and can lead to ignoring the risks associated with hands-free mobile usage—which Rosenberger exposes through means such as a postphenomenological analysis—in favor of waiting for the technological fix provided by the advent of autonomous vehicles on the roadways. Giving in to "spectatorial utopianism," he concludes, is a way of decreasing highway safety rather than the other way around.

Chapter 7 expands the focus of this volume beyond ethical issues associated with individual self-driving vehicles. Shane Epting's interest in "Automated Vehicles and Environmental Justice: Addressing the Challenges Ahead" lies in how transportation-related environmental justice outcomes for residents of marginalized communities could be improved through the adoption of automated vehicles. Getting to this end requires resisting "armchair philosophizing" about environmental justice in the abstract in favor of paying attention to how members of these communities will use automated (including electric) vehicles in the course of their everyday working and family lives, which, as the environmental justice theorist Robert Figueroa emphasizes, are embedded in particular neighborhoods, cultures, and histories whose destruction to make way for new highways constitutes environmental injustice. Urban city planners, Epting argues, need to meet the challenge of bringing the voices of marginalized citizens into planning conversations so the introduction of automated vehicles will intentionally lead to environmentally just outcomes and not just "trickle down" to them.

With chapter 8, "Vehicles of Change: The Role of Public Sector Regulation on the Road to Autonomous Vehicles," Patrick Schmidt and Jeremy Carp invite the reader to also explore the need for the public's involvement in decisions regarding transformative technologies, of which for them autonomous vehicles serve as an illustration. Their contribution's title offers a pointer for how this might be done. If "vehicle" simply means "autonomous vehicle technology," then US regulatory approaches tend to veer conservative. Regulators face a catch-22. Fewer regulations can bring about a disconnect between technology and the law, while more regulations run the risk of becoming outdated. Schmidt and Carp imaginatively propose that "vehicles" can also stand for "public regulations." They suggest that law and technology view themselves as "evolutionary partners" in building resilient, adaptive systems that would involve the public in their design.

Also breaking away from the idea that the only ethical considerations to be attended to with regard to autonomous vehicles stem from individual vehicles themselves are Joseph Herkert, Jason Borenstein, and Keith W. Miller, in "Planes, Trains, and Flying Taxis: Ethics and the Lure of Autonomous Vehicles" (chapter 9). The authors look at ethical issues emerging from large-scale sociotechnical transportation systems involving interactions extending beyond cars and personal trucks to a range of other autonomous vehicles whose development is planned or in progress, from buses, shuttles, and delivery drones to commercial trucks, planes, and cargo ships. Herkert, Borenstein, and Miller argue that there are ethical issues connected to these developments that strike deeply into people's everyday lives, such as being pushed out of the labor force, social justice, and accessibility for all users. They recommend that the focus of the development and regulation of autonomous vehicles be put on mass community-based transportation and that workforce loss issues be dealt with at the same time as AVs are being rolled out.

Wrapping up this volume on ethical issues connected to autonomous vehicles in everyday settings are Ike Kamhof and Tsjalling Swierstra. Their "Experiencing the Future: A Postphenomenological Exploration of Autonomous Vehicles" (chapter 10) begins by questioning the prominence of "hard impacts" in policy making for self-driving cars. They pose, from a postphenomenological viewpoint, that emerging technologies have the power to transform lived experience in many ways, and that the resulting "soft impacts"—ones that are *qualitative, ethically ambiguous, and coproduced by social and technical factors*—need ethical analysis and attention, as these are the changes that will play a real part in people's lives. To further support their point, Kamhof and Swierstra develop four phenomenological "anecdotes" designed to show how autonomous vehicles can work to shape lives at the *level of lived, mundane experience.* It is possible, they suggest, to have a phenomenological experience of the future; moving forward, what can be learned from these anecdotes can be useful to engineers and designers as well as to policymakers and the general public alike in anticipating ways that the soft impacts of autonomous vehicles might work to change lived experience so they can be weighed as these vehicles are still in development, rather than waiting until afterward when they might be too hard to undo.

THE ETHICS OF TECHNOLOGICAL CHANGE

The contributions to this volume share a common focus on some of the ethical issues associated with one example of technological change: namely, the development of autonomous vehicles. In line with the full subtitle of this volume, *Autonomous Vehicles and the Ethics of Technological Change*, we could ask what could be learned from the insights offered in these contributions regarding this one particular example of technological change when it comes to imagining and identifying the ethical issues involved with technological change in general, no matter what the specific innovation might be whose development is already underway or poised on the developmental horizon.

It is not hard to see why it matters to envision what an ethics of technological change might look like. As Philip Brey has noted in the course of his own contribution to this emerging field of inquiry (2012), the sooner it is possible to reliably identify what ethical concerns a particular emerging technology might pose, the sooner these challenges might be able to be addressed. This is particularly important, we might add, when it comes to emerging technologies such as autonomous vehicles that are transformative innovations whose possible impacts carry with them the potential for rippling and cascading through multiple sectors of many societies worldwide, and so shaping the lives of users and nonusers as well.

While interest in an ethics of technological change is building, to repeat what was said above, it is still itself an emerging field of inquiry. Among the contributions propelling the growth of the field are Brey's own anticipatory technology ethics (2012), ethical technology assessment as laid out by Palm and Hansson (2006), and Ibo van de Poel's theory that technologies in the making are best viewed as social experiments (2016). At the risk of oversimplifying greatly, the first two approaches involve a checklist of key values such as privacy, sustainability, and justice that anyone involved in the development of a particular technology ought to be cognizant of. For Palm and Hansson, the values checklist can be used as a basis for ongoing conversations between ethicists and those involved in the design process in order to mitigate potential negative impacts in advance, and so play the role of an "early warning system" (p. 551). Brey's approach, which includes an extensive checklist of values with a special emphasis on harms to be avoided, human rights, distributive justice, along with well-being and the common good, is also intended to be used in what he calls the "design feedback" stage (p. 12). Drawing on widely accepted principles of biomedical ethics—nonmalfeasance, justice, beneficence, and respect for autonomy—as grounding principles, van de Poel connects them to a set of conditions that can be used to judge whether the introduction of a new technology is an acceptable social experiment, including whether it looks as though social benefits will follow from the experiment and how easy it is to stop the experiment once it gets underway (p. 680).

While there are significant differences among these three ways of approaching an ethics of technological change, they share a similarity in that they are focused on the innovative aspect of a particular artifact as it goes through the beta stages of design and entry into social life. It is also possible to construct an ethics of technological change that would take social life itself as the starting point for analysis when assessing the possible negative impacts of a novel artifact and its features. Such an ethics of technological change would center reflection on everyday environments, with its components of the trustworthiness of lived experience, embedded narratives, and culturally based communicative conventions.

What might these considerations look like? The following is more of an initial sketch than a finished product, and in that sense it is closer to being a checklist than a holistic ethical framework. While it is intended to represent a different viewpoint to the perspectives discussed earlier, it is not intended to be an alternative to them but to be complementary instead, as whether a particular technology taking shape will perpetrate discriminatory attitudes, or social inequalities such as a disproportional distribution of benefits and risks among members of marginalized groups, or raise concerns about the protection of privacy, or diminish freedom of choice and autonomy, or threaten human rights, are all critical issues that need to be brought to the forefront during the design process.[5] The approach suggested here to an ethics of technological change could be called vernacular value visioning (VVV). It is worth reflecting, in the early stages of its development, whether and to what degree a particular technology in the works might lead to:

- *Value-shrinkage.* When a particular interpretation of an everyday value becomes the standard due to the introduction of a particular technology, other interpretations and narratives may be threatened and may be difficult to restore once the technology has entered widespread use. So, for example, contemporary media technologies have led to a particular view of what counts as beauty, to the exclusion of other, legitimate perspectives.
- *Value-colonization.* Just as value-shrinkage involves the spread of a particular value, value-colonization occurs when the main or subsidiary ethical values emphasized by a new technology—safety and efficiency, for example, in the case of autonomous vehicles—spread out into and infiltrate certain everyday dimensions of experience and human interaction to the detriment of these areas.

- *Value-leapfrogging.* Connected to value-colonization is the phenomenon of value-leapfrogging or bypassing, which is connected to the role of validation in systems design. As Luciano Floridi has observed, the step of validation demands that any proposed technological system is vetted as being the right solution to a problem by being *necessary* (as contrasted with other possible solutions for the problem), by *fitting proportionately* to the problem to be solved, and by being *confined to* addressing *just* this particular problem (Floridi 2020). Once this stage is met one can go on to figure out what the right way might be for designing this system. A basic assumption behind these activities is that the solution to the problem would be a technological one. The more preliminary question of whether it would be possible to draw down from the resources of lived experience and symbolic means of communication in order to find the solution could be drawn from the resources of lived experiences and symbolic forms of communication is one that gets leapfrogged over.
- *Value-hijacking.* Let us define value-hijacking as a movement that takes place when a value that is central to established practices of everyday life gets undermined by an emerging technology and placed in doubt. A possible example would be trusting the perspective of a GPS more than one's own navigational knowledge, even when the GPS appears to be incorrect; more broadly speaking, it would mean putting more trust into an algorithm than ourselves, our own culturally informed experiences, and the validity of our gaze.
- *Value-gapping.* To what degree despite early stage interventions in the design process might the proposed technology in question work to widen the gap between its integration into social life and the capacity to respond to it ethically?
- *Value-deleveraging.* Will the emerging technology be able to leverage more opportunities for human beings to live more attentive and imaginative lives, or might they serve to reduce them instead? If it can do the former, it could serve to make societies less vulnerable and more resistant to certain technological imaginaries, such as ones that stress the importance of gamification and infotainment.
- *Value-disruption.* Value-disruption is closely tied to value-leapfrogging. It can be said to occur when a more disruptive solution to a problem is opted for rather than one that is less disruptive. At least in principle, a less disruptive solution would arguably be more affirming of everyday practices and social conventions.

This volume, including this "preview" as to what an ethics of technological change thought of as VVV might look like, has come together at a point in time when the answer to the question *Will there be a future when autonomous vehicles will dominate the roadways?* is very much up in the air. When the development of autonomous vehicles first began to make media headlines, closing the distance between the present and the future was seen as accelerated and the future was often predicted to be just around the corner. In 2016, for instance, the Ford Motor Company affirmed it would have driverless vehicles at Level 4 autonomy without steering wheels or brakes to be used for ride-hailing purposes on the highways by 2021 (Della Cavo 2016). In the same year, the cofounder of the ride-hailing company Lyft claimed that before 2021 the majority of Lyft rides would be in fully autonomous vehicles (Zimmer 2016). Even earlier, Nissan announced it would be able to market fully self-driving vehicles to the public starting in 2020 (White 2013).

With the COVID-19 pandemic, road-testing of driverless cars was suspended. The pandemic introduced a deep pause in these ambitious narratives and pushed the arrival of substantial numbers of fully autonomous vehicles out into the indefinite future. But the pandemic was neither the only nor the primary factor involved in this slowdown. "Teaching" autonomous vehicles how to drive has largely been dependent on machine learning, but that has turned out to have more limitations than initially thought. One novel plan to speed up testing driverless vehicles is to thoroughly train AI drivers in virtual environments using synthetic data before testing is done in real-world environments (Heaven 2022). Still, just what kind of role AI will play in the future in the development of autonomous vehicles remains an open question (Mims 2021).

To be sure, while this technical challenge is substantial, there is a substantial cultural challenge that needs to be dealt with as well and that in some sense could be said to be the main challenge of test-driving the future when it comes to autonomous vehicles. That challenge—one that might not be met until the activity of engineering is reenvisioned as being concerned with morality at its very core (Doorn et al. 2021, 739)—is to turn the hope that reflections on ethical values would be thoroughly incorporated in the technology design process into a concrete reality. That would involve a shift within the culture of engineering, away from thinking of itself as a discipline grounded in making progress in problem solving and opening it up to a broader self-vision. How this might be accomplished lies well beyond the scope of this volume. For the time being, given the "reset" that the production of autonomous vehicles is now going through, it is hoped that the contributions to this volume might be able to enjoy some impact on their design space.

That leads to one final note in wrapping up this chapter. While the rollout of autonomous vehicles has been delayed, the opposite has been the case with philosophical reflection on the ethics of autonomous vehicles. This has come a long way from its beginnings in thinking about the trolley problem. In this rare state of affairs, the ethics-technology gap has been flipped, with ethical reflection running ahead of technological advances. And that can be seen as a good thing indeed.

NOTES

1. A Google search for "autonomous vehicles" in early 2022 turned up well over a billion and a half results. "Self-driving vehicles" had nearly 970 million results, and there were close to thirty million results for "driverless vehicles."

2. In the course of wondering whether autonomous vehicles should be programmed to make "lite" driving infractions as is common for humans to make while out for a drive (such as going through a yellow light before it turns red), Nyholm and Smids (2020) cite Raj Rajkumar, director of the Carnegie Mellon University automated driving lab, as saying (p. 338) that for the time being the automated vehicles coming from the lab would be programmed to obey the rules of the road. This certainly reduces programming complexity and sets expectations for other drivers. Still, the consequences of such a decision could make the behavior of an autonomous vehicle more predictable and so more susceptible to cyberattack. The end result could be to decrease rather than increase vehicular safety.

3. These numbers are preliminary estimates. We might trust the accuracy of these numbers but be skeptical of their meaning. The fact that roadways were less congested during the COVID-19 pandemic might have led to drivers taking more risks, resulting in an increase in the number of fatal accidents.

4. Himmelreich is skeptical that trolley problems are central to the ethical issues associated with automated vehicles and gauges that the ethical focus needs to shift to mundane mobility situations, such as approaching a crosswalk that a pedestrian is about to enter. Himmelreich's focus on the everyday is aligned with the emphasis of this volume. In particular, he calls attention to how such mundane situations illustrate Moravec's paradox highlighting that what we find easy to do is hard for an AI to execute (2018, 11–12). He does though understand this ease as being a matter of intuitive know-how rather than tying the ease to the entanglement of lived experience with social conventions.

5. For an example of how autonomous cars might reinforce discriminatory attitudes, see Maya Indira Ganesh (2016) as referenced in Fiona McDermott (2020).

REFERENCES

Bonnefon, Jean-François, Azim Shariff, and Iyad Rahwan. 2016. "The Social Dilemma of Autonomous Vehicles." *Science* 352 (6293): 1573–76.

Brey, Philip A. E. 2012. "Anticipatory Ethics for Emerging Technology." *Nanoethics* 6: 1–13.

Capurro, Rafael. 2019. "Enculturating Algorithms." *Nanoethics* 13: 131–37.
Davnall, Rebecca. 2020. "Solving the Single-Vehicle Self-Driving Car Trolley Problem Using Risk Theory and Vehicle Dynamics." *Science and Engineering Ethics* 26 (1): 431–49.
Della Cavo, Marco. 2016. "Ford Promises Driverless Cars by 2021." *USA Today*, August 16. https://www.usatoday.com/story/tech/news/2016/08/16/ford-promises-driverless-transport-2021/88826072/.
Doorn, Neelke, Diane P. Michelfelder, Elise Barella, Terry Bristil, Francien Dechesne, Albrecht Fritzsche, Gearold Johnson, Michael Poznic, Wade L. Ribison, Barbara Sain, Taylor Stone, Tonatiuh Rodriguez-Nikl, Steven Umbrella, Pieter E. Vermaas, Richard L. Wilson. 2021. "Reimagining the Future of Engineering." In *The Routledge Handbook of the Philosophy of Engineering*, edited by Diane P. Michelfelder and Neelke Doorn, 736–44. New York: Routledge.
Floridi, Luciano. 2020. "Mind the App: Considerations on the Ethical Risks of Covid-19 Apps." *Philosophy and Technology.* DOI 10.1007/s13347-020-00408-5.
Fritzsche, Albrecht. 2021. "The Artefact on Stage: Object Theatre and Philosophy of Engineering and Technology." In *Engineering and Philosophy: Reimagining Technology and Social Progress*, edited by Zachary Pirtle, David Tomblin, and Guru Madhavan, 309–22. Springer Nature Switzerland.
Ganesh, Maya Indira. 2016. "Cities After Cars, Places After Data." *The Society Pages: Cyborgology*, December 1. https://thesocietypages.org/cyborgology/2016/12/01/cities-after-cars-places-after-data/.
Hansson, Sven Ove, Matts-Åke Belin, and Björn Lundgren. 2021. "Self-Driving Vehicles—An Ethical Overview." *Philosophy and Technology* 34 (4): 1383–408.
Heaven, Will Douglas. 2022. "This Super-Realistic Virtual World Is a Driving School for AI." *MIT Technology Review*, February 18. https://www.technologyreview.com/2022/02/18/1045784/simulation-virtual-world-driverless-car-autonomous-vehicle-school-ai-cruise-waabi/.
Hicks, Daniel J. 2018. "The Safety of Autonomous Vehicles: Lessons from Philosophy of Science." *IEEE Technology & Society Magazine* 37 (1): 62–69.
Himmelreich, Johannes. 2018. "Never Mind the Trolley: The Ethics of Autonomous Vehicles in Mundane Situations." *Ethical Theory and Moral Practice* 21: 669–84.
Hui, Yuk. 2016. *The Question Concerning Technology in China: An Essay in Cosmotechnics.* Falmouth, UK: Urbanomic Media.
Lawlor, Rob. 2022. "The Ethics of Automated Vehicles: Why Self-Driving Cars Should Not Swerve in Dilemma Cases." *Res Publica* 28: 193–216.
Leben, Derek. 2017. "A Rawlsian Algorithm for Autonomous Vehicles." *Ethics and Information Technology* 19 (2): 107–15.
Lin, Patrick. 2016. "Why Ethics Matters for Autonomous Cars." In *Autonomous Driving: Technical, Legal, and Social Aspects*, edited by Markus Maurer, J. Christian Gerdes, Barbara Lenz, and Hermann Winner, 69–85. Springer.
Lundgren, Björn. 2021. "Safety Requirements vs. Crashing Ethically: What Matters Most for Policies on Autonomous Vehicles." *AI & Society* 36: 405–15.
McDermott, Fiona. 2020. "New Sensorial Vehicles: Navigating Critical Understandings of Autonomous Futures." In *Architecture and the Smart City*, edited by Sergio M.

Figueiredo, Sukanya Krishnamurthy, and Torsten Schroeder, 247–56. New York: Routledge.

Milano, Silvio, Mariarosario Taddeo, and Luciano Floridi. 2020. "Recommender Systems and Their Ethical Challenges." *AI and Society* 35: 957–67.

Mittelstadt, Brent, Patrick Allo, Mariarosaria Taddeo, Sandra Wachter, and Luciano Floridi. 2016. "The Ethics of Algorithms: Mapping the Debate." *Big Data and Society* July–September: 1–21.

Mims, Christopher. 2021. "What If Truly Autonomous Vehicles Never Arrive?" *The Wall Street Journal*, Saturday–Sunday, June 5–6, p. B4.

National Highway Traffic Safety Administration. February 2022. "Early Estimate of Motor Vehicle Traffic Fatalities for the First 9 Months (January–September) of 2021." Report No DOT 813 240.

National Highway Traffic Safety Administration. May 2021. "Early Estimate of Motor Vehicle Traffic Fatalities in 2020." Report No DOT 45 813 115.

Nyholm, Sven, and Jilles Smids. 2020. "Automated Cars Meet Human Drivers: Responsible Human-Robot Coordination and the Ethics of Mixed Traffic." *Ethics and Information Technology* 22: 335–44.

Palm, Elin, and Sven Ove Hansson. 2006. "The Case for Ethical Technology Assessment." *Technological Forecasting and Social Change* 73: 553–58.

Sparrow, Robert, and Mark Howard. 2017. "When Human Beings Are like Drunk Robots: Driverless Vehicles, Ethics, and the Future of Transport." *Transport Research Part C: Emerging Technologies* 80: 206–15.

Van de Poel, Ibo. 2016. "An Ethical Framework for Evaluating Experimental Technology." *Science and Engineering Ethics* 22 (3): 667–86.

White, Joseph. 2013. "Nissan Pledges 'Multiple' Self-Driving Cars." *The Wall Street Journal*, August 27. https://www.wsj.com/articles/nissan-expects-to-market-selfdriving-cars-by-2020-1377623274.

Zimmer, John. 2016. "The Third Transportation Revolution." *Medium*, September 18. https://medium.com/@johnzimmer/the-third-transportation-revolution-27860f05fa91#.f1csicx9o.

Chapter 2

A Postphenomenology of Electric, Self-Driving, and Shared Vehicles

Galit Wellner

A history of the automobile may start with a horse-drawn carriage that obeys the rules of Newtonian physics. This is the foundation of the dream of a vehicle that will carry passengers easily, quietly, and effortlessly. The next stage starts in the late nineteenth century with the invention of the horseless carriage—that is, a car powered by fuel. This car is still subject to Newton's theory but also to the more complex theory of thermodynamics. The autonomous vehicle (AV) that is much discussed today is the contemporary manifestation of this dream. The realization of the dream is gradual, yet recently it seems to have accelerated. This acceleration becomes visible when compared to the car's development throughout most of the twentieth century, when the basic operation of a car did not change significantly. That is not to say cars did not change at all. On the contrary, their silhouettes changed toward more aerodynamic shapes, their safety increased due to the addition of safety belts and other protective measures, automatic gears simplified the operation of the vehicle, and even a modest act like opening the side windows changed, for example by the implementation of electric mechanisms. However, the basics remained—a driver's seat in front of a steering wheel and a dashboard, passenger seats next to the driver and behind, and above all the principle that the driver must be a conscious human being whose attention should be paid to the act of driving (cf. Irwin 2014; Michelfelder 2014; Wellner 2014).

In the twenty-first century this paradigm has been shifting toward AVs. As this chapter is being written, this shift is still underway, and its next major steps can only be speculated. Yet it is important to understand the directions of development in order to formulate policies aimed at increasing the benefits and reducing the harms, dealing with questions such as where passengers can embark and disembark, or who receives priority on the road. To understand the evolution of AVs, it is important to understand the diverse technologies that comprise it. This chapter is an attempt to unpack AVs and comprehend the various effects of these technologies. AV subtechnologies are divided here into three technological families: electric engines, self-driving mechanisms, and shared ride algorithms. Whereas the latter two are easily associated with future AVs, the electric engine is regarded as more general. It is conceived by many as the current motor of vehicles, and hence it is considered here as part of the AV, although today most electric cars are not autonomous. Each technological family is comprised of technologies that work together or offer alternative solutions to obtain a certain way of moving from point A to point B, each covering a different aspect of such a movement. Sometimes these families cooperate with each other, like electric autonomous vehicles, but this is not a must. A car can be autonomous and fueled by gasoline, or vice versa.

Orienting ourselves toward the responsible development and implementation of AVs, the policies for regulating AVs should not just be focused on the technologies but rather should take into account the interactions with human users and nonusers. In this chapter I wish to contribute to future regulation by analyzing the entanglements among humans, their technologies, and their surroundings, as well as upcoming technological changes and regulatory regimes (for example, sensors that detect the passenger's eyes, or regulation expanding the principles of equality to the shared vehicle). To do so, I employ a theory and methodology known as postphenomenology, a branch of the philosophy of technology that studies the coshaping between humans and technologies. Such studies are performed through a formula composed of three ingredients: I, technology, and world. Postphenomenology playfully adds parentheses and arrows to denote the various relations between the ingredients. Originally Don Ihde devised four relations (Ihde 1990), which will be described in the second section. Next Peter-Paul Verbeek (Verbeek 2008a) developed three additional relations under the umbrella of technological intentionality, to be detailed in the third section. The most recent development is an attempt to rethink human intentionality when artificial intelligence (AI) technologies express intense technological intentionality. This will be the focus of the fourth section. Last, I will show how the "world" component of the postphenomenological formula should gain an increased visibility, especially in light of the environmental crisis we are facing.

To each developmental stage of the theory of postphenomenology, I match a specific type of car and its associated technology: electric vehicles (and the electric engine), self-driving cars (and their sensors), and shared ride vehicles (and the algorithm) (see table 2.1).

Consequently, this chapter moves in two parallel and interrelated evolutionary paths: one that describes the developments in the postphenomenological

Table 2.1. AV Technologies and Their Matching Postphenomenological Frameworks

Car Type	Core Technology	Postphenomenological Framework
Electric vehicle	Electric engine	Four basic relations
Self-driving vehicle	Sensors	Technological intentionality
Shared ride vehicle	Algorithms	Relegation

formula of "I – technology – world," and the other focused on recent developments in car technologies.

THE ELECTRIC ENGINE AND CLASSICAL POSTPHENOMENOLOGY

Electrical vehicles are already present in the roads of many countries, and charging stations are mushrooming in parking lots, streets, and even gas stations. Basically, these cars do not consume fossil fuels but instead charge from the power grid. This family of technologies includes batteries, charging methods, and above all a new engine. There are still some major challenges ahead, like the number of kilometers per charging or the recycling of used batteries.

Let's analyze the electric car's relations with its drivers through the work of Don Ihde (1979; 1990). He unpacks Martin Heidegger's dichotomy of ready-to-hand vs. present-at-hand into four basic types of relations among humans, technologies, and the world. These relations are universal in the sense that they can be found in ancient tools, modern machines, or contemporary digital technologies. The specific manifestations change, of course, but the basic principles remain.

The first postphenomenological relation, *embodiment relations*, refers to a situation in which technologies change the user's body scheme and become part of it. One of the classical examples of embodiment relations is driving a car, a situation in which the driver knows where she and her car can pass through a city's narrow streets or avoid hitting other cars while parking. This relation is represented by the following permutation of the postphenomenological formula: (I – technology) → world. The parentheses indicate that the user-driver and technology-car operate as a temporarily unified unit in the world. Embodiment relations remain applicable for the electric vehicle, as long as such a vehicle maintains today's shapes and principles of automotive operation.

The second relation is termed *hermeneutic* because it involves reading and interpreting the world. The formula denoting this relation is: I → (technology – world). Here the technology and the world form a unity for the experiencing

I. The user reads the world through the mediation of the technology, so that technology and the world function together as one unit to be interpreted. Readings can be of text but also of graphic representations, as in the case of a dashboard in which the indications for the car's states can be presented numerically, on a scale, or even as a simple LED green or red light. All these indications require some interpretation, such as understanding the meaning of a red color or realizing what number (or what point on a scale) represents a cause for concern. Hermeneutic relations occur not only on the visual level but also on the auditory. The driver needs not only to look at the dashboard but also to attend to the roar of the engine and the swoosh of the wheels as they touch the road. An electric vehicle poses a significant change in the hermeneutic relations, as electric engines are silent. This fact leads to a new type of interaction in which the engine's conditions are not transmitted spontaneously to the hearing driver but rather should be predesigned via digital interfaces, which can be oral or visual.

So far, the parenthesis indicated how a technology is regarded as part of either the "I" or the "world." In the third and the fourth relations the role of the parenthesis changes, and instead of symbolizing a temporary unity, it stands for a kind of bracketing. Thus, in alterity relations in which the user interacts with the technology as if it were a quasi-other, the postphenomenological formula represents that the world withdraws to the background: I → technology (– world). Many car fans are familiar with alterity relations, as they give a name to their cars and take care of them. Electric vehicles are no exception. The concept of alterity relations helps us to model the driver's responsibility to make sure the car has enough energy to complete the route and arrive at the destination, where it can be recharged. In this sense, the car keeps the pattern of relations between a horseman and a carriage led by horses, followed by the need to make sure there is enough gasoline in the car's fuel tank. Electric vehicles can be regarded as simply replacing the fuel tank with a battery that needs to be charged.

Last, in background relations the technology withdraws to the background, working unnoticed. It is represented by the following permutation of the postphenomenological formula: I → (technology –) world. A car can be in the background, as car rental agencies in Japan have recently discovered.[1] They noticed that many cars are returned with zero mileage. When asking their customers, it turned out that they were using the rented cars for various purposes: to store items when nearby lockers were full, to charge their cellphones, to sleep, or to eat lunch quietly. In Japan the rented cars withdrew to the background in order to create an environment according to the user's needs. Electric vehicles can basically maintain these uses. While being driven, they follow a longtime trajectory of an effort to keep the car in the background, silent.

In short, the four postphenomenological relations are universal and remain as relevant and applicable for electric vehicles as they were for gasoline-powered cars. However, some modifications are required to accommodate the fact that the engine is silent. The technologies for self-driving and shared rides, however, call for new permutations of the formula, beyond the four presented by Ihde in 1979 and 1990.

SELF-DRIVING VEHICLES, SENSORS, AND TECHNOLOGICAL INTENTIONALITY

The dream has been around for decades (and maybe even centuries)—a car that can drive by itself, eliminating the need for a human driver. Such a car should be able to navigate by itself, accelerate and decelerate according to traffic conditions and rules, keep a safe distance from other cars, etc. This requires the addition of sensors and cameras to a car so that it can "see" the environment in real time. The sensors' output is analyzed by algorithms that turn the "seeing" into an "understanding" of the road and the traffic conditions. They direct and control the car's movement second by second (and even millisecond by millisecond!). In addition, the vehicle can be connected to the internet to obtain maps or updates on the traffic, but this is not a must.

The main challenges are located (at least at this stage) in the immediate detection of external conditions such as pedestrians unexpectedly moving onto the road or unplanned roadblocks like a stuck car. A significant amount of discussion has been devoted to a thought experiment known as the trolley problem in which a driver has to choose whom to hit, depending on the number of people in each group and their identity (for an overview, see Himmelreich 2018). Although this thought experiment has gained a lot of traction, in real life such situations are extremely rare. Therefore the ethical discussion should not focus on the trolley problem but instead should deal with more practical ethical questions. For example, we can ask whether passenger safety should have priority over the safety of pedestrians or bikers. But a wider perspective would allow us to switch from questions regarding death to questions regarding life—such as what it is like to live with these technologies if you are a member of a minority group or to consider their impact on the environment and on future generations (see Epting 2018; also Santoni de Sio 2017; Himmelreich 2018; Borenstein, Herkert, and Miller 2019).

In this section, though, I shall focus on the user's perspective, according to which the self-driving car has to bring the user-passenger from point A to point B. It is a full automatization of the driving process in which the vehicle and the user share the same goal. To analyze this joint goal experience, I employ the developments to postphenomenology offered by Peter-Paul Verbeek (2008a).

Verbeek expanded upon the four basic postphenomenological relations originally formulated by Ihde by introducing the notion of technological intentionality. Similarly to human intentionality, "'technological intentionality' here needs to be understood as the specific ways in which specific technologies can be directed at specific aspects of reality" (2008a, 392). Technological intentionality is interpreted as the ability of technologies to form intentions so that they enable the users to do things that could hardly be done without such technologies. There is obviously a difference between human intentionality and technological intentionality: "Even though artifacts evidently cannot form intentions entirely on their own, . . . because of their lack of consciousness, their mediating roles cannot be entirely reduced to the intentions of their designers and users either" (2008b, 95). For Verbeek, technological intentionality supports human intentionality so that "when mediating the relations between humans and reality, artifacts help to constitute both the objects in reality that are experienced or acted upon and the subjects that are experiencing and acting" (95).

Based on the notion of technological intentionality, Verbeek adds to the postphenomenological relations new permutations, one of which he terms "composite intentionality." This permutation represents situations in which not only human beings have intentionality but so do the technological artifacts that they are using. In composite intentionality, the "directedness" of a technology is added to the human intentionality. The intentionality is added hermeneutically; that is to say, there are new ways in which the technology "reads" the world. This structure of double intentionality is represented in the postphenomenological formula by two arrows: I → (technology → world). This is a modification of the hermeneutic relations' formula in which the technology and the world are connected by a dash. The arrow indicates intentionality, and when pointing *from* the "technology" it is no longer associated solely with the human "I." It models situations in which the technology is imbued with some independence and ability to decide, to take direction (see also Wellner 2017; 2018; 2020).

Composite intentionality can model how self-driving vehicles and their passengers get from point A to point B. The user defines her starting point and destination, and the self-driving vehicle works out the details of the way. This is represented by the left arrow that departs from the "I." The rest—"technology" and "world"—are conceived as a unit that functions autonomously. The right arrow that connects "technology" and "world" represents the directedness of the technology. Here the "technology" is a group of technologies that functions together as a system. First the system calculates the route and then in real time it navigates to the destination while micro-navigating through the other cars, road conditions, and pedestrians. The self-driving system and the passenger-user share the same set of intentions, but each has a different way of expressing them. They both want to depart from a certain place and go to another, they both want to get there as soon as possible, and they both wish to avoid hitting pedestrians or other vehicles on the way.

This is, of course, an idealized picture. In practice, the system is likely to take into account additional parameters. Think, for example, about the possibility that a self-driving car will attempt to maximize the passengers' exposure to certain businesses, and so will travel along certain routes where those businesses can be found or where certain street advertisements have been put. The user becomes passive, in contradiction to the formula representing composite intentionality that assumes a certain flow going from the human to the technological and the world. The relations known as "composite intentionality" and the related postphenomenological formula hardly fit situations in which we "give up" our intentionality, delegate our decision making to a technology, and let the technology lead, decide, and operate. Composite intentionality cannot accommodate situations in which the human intentionality "withdraws." A new type of relations is required. This is what I termed (Wellner 2020) relegation, and I will discuss this relation in the context of AV in the next section.

SHARED RIDES AND RELEGATION

The third family of AV technologies is the shared rides that have been slowly penetrating metropolitan areas. These technologies combine existing means of transportation with algorithms and apps that manage them. The algorithms can manage a fleet of vans each functioning like a taxi for multiple passengers with different departures points and destinations. The algorithms can also manage the sharing of various transportation means such as cars, bicycles, or electric scooters. Each vehicle can be rented for a short period, measured in minutes or hours. Instead of owning such a vehicle, one can rent it in one location and return it to any other point in the (usually metropolitan) area. The apps facilitate the rental and interact directly with the user. The background algorithm manages the number of vehicles in various parts of the city.

The focus of this section is on the first type of sharing performed for taxi (or minibus) rides through a mobile app. Basically this family of technologies enables taxi passengers to travel with strangers, whose destination has a proximity with the others' paths and destinations.[2] The core technology here is an algorithm that matches taxis to passengers. From the user's perspective, shared rides open the possibility for customized public transportation, saving the need to switch buses or other means of mass transportation. From a policy perspective, shared taxi rides are intended to alleviate road congestion, as well as to reduce the carbon footprint of the transportation sector, together with other technologies like electric vehicles discussed in the previous sections.[3] It should do so by encouraging more people to use public transportation and reduce the usage of private cars, whether autonomous or not (cf. Epting 2018). From the software developers' perspective, the main challenge is how to optimize the route, based on the departure points and destinations of the passengers, traffic conditions, and the energy required by the car to reach the final destination (which turns out to be crucial for the current battery life of electric vehicles).

Returning to the user's perspective, a passenger in a shared ride can basically control only the starting and destination points. In the simplest form, the shared rides' algorithms calculate how to get from point A to point B while trying to find a common route for each passenger. These algorithms direct the drivers of the shared taxis or the autonomous driving system to go through certain streets and optimize the path according to traffic conditions as they may be from time to time. By calculating the common route, the algorithms exhibit an intentionality that is "composite," that is, combined of the user's human intentionality and the software's technological intentionality.

Supporting and enhancing human intentionality is but one option for shared ride algorithms. In practice, when these algorithms calculate the route, it is not necessarily optimized for the needs of any passenger. Instead they take into account the interests of all passengers, that is, where to be picked up and taken to. Preferably these places should be near to the routes of other passengers. Sometimes the technological intentionality of the algorithm takes over and becomes more dominant. This happens, for example, when the algorithm "decides" to pick up a new passenger located off the route(s) of those who are already in the car.[4] This pickup contributes to the revenues of the operating company but is at the expense of the other passengers, as they will reach their destinations later than originally planned.

In some systems, even the departure and end points of the journey are not fully controlled by the user, as the algorithm requires the passenger to depart from what it considers to be the closest bus station, and the same for the destination. Additionally, in most systems the number of passengers and the identity of the other travelers are controlled by the algorithm (which can be sensitive from the perspective of a young woman embarking on a late night shared ride taxi that is full of men).

Some of this might sound like the dystopian descriptions of the machines in the early days of the Industrial Revolution. But the situation here is different. While tools were perceived by Karl Marx to be in the control of humans, and machines to be controlling of humans, Gilles Deleuze and Felix

Guattari (1987) argue that humans and digital technologies are pieces of the same system. The human and the technological artifact operate in a complex mechanism in which they are no more than components (see Wellner 2022). They call it "the third age of technical machines."[5] They elaborate:

> It is the reinvention of a machine of which human beings are constituent parts, instead of subjected workers or users. If motorized machines constituted the second age of technical machines, cybernetic and informational machines form a third age that reconstructs a generalized regime of subjection . . . *the relation between human and machine is based on internal, mutual communication, and no longer on usage or action.* (Deleuze and Guattari, 1987, 458; emphasis added)

More radically than modern machines, digital technologies are not necessarily at the service of the human. These technologies call for a redefinition of their relations to the human and it is difficult to determine who and what is superior. AV technologies exemplify this new kind of relations. Who decides which way to go? Who determines the identity of the other passengers?

Let us represent this role of the algorithm in the postphenomenological formula. In the previous section I discussed the addition of an arrow that connects technology and the world. In order to represent situations in which human users are "subjected" to the technologies they use, we need to reverse the human intentionality arrow in the postphenomenological formula so that it points *to* the human and not *from* the human: I ← (technology → world).

In this permutation, the human intentionality "withdraws" and the technological intentionality "takes over." It reflects situations in which technologies control the world as well as the users. It is an extreme form of technological intentionality, which "forces" the user to obey the technology. Drivers that go in loops because they obey the navigation app's instructions are just one example (Wellner 2020). It is an extreme delegation, compared to Bruno Latour's door example (Latour 1992), in which the human doorman was replaced by an automatic door. In Latour's example, closing the door is "delegated" to an automatic mechanism. Here, however, the human remains in the picture, but only as a passive participant. I term this situation "relegation" (Wellner 2020) because the human intentionality is assigned to an inferior position. The algorithms "downgrade" the human intentionality. In other words, not only there is no symmetry, control now resides on the side of the nonhumans.

We have seen this type of technological intentionality in the Cambridge Analytica affair, when a company used Facebook's profile system and the advertisement mechanisms to shape the users' political views and voting. The users experienced Facebook as shaping their wishes, desires, and fears.

It should be noted that this experience was not spontaneous but rather the result of Facebook's efforts (Zuboff 2019). An analysis that remains faithful to human intentionality would regard the role of the technological platform as a means to an end and would seek to reveal the people operating the software. But the emergence of AI urges us to model situations that are not driven by human intentionality. Human intentionality is subjugated. This is also what Shoshana Zuboff (2019) revealed when she studied the "economies of action" of Facebook, Google, and Pokemon Go. She describes how these companies treat their users as data generators that can and should be manipulated through "behavioral modifications" without their knowledge or consent. Thus, intentionality can be imputed to technologies and the notion of relegation becomes central to the relational analysis.

Yet relegation can have a positive impact. Think of an autonomous shared taxi whose passengers leave the car dirty or harass other passengers. They allow themselves to behave in such ways in the absence of a driver in this space. Sensors installed inside the car can create a "third person's gaze" effect and reduce the probability of such unwanted events. Such situations may call for a further development of the postphenomenological formula in which the right arrow connecting "technology" and "world" is reversed to model the intentionality of the environment: I → (technology ← world).

Relegation—in its two permutations—helps us to model the role of algorithms that decide on which route we will go, how fast we are going to arrive to our destination, who else will be with us in the car, and how they may behave.

SUMMARY AND CONCLUSIONS: A NEW ROLE FOR THE WORLD

In this chapter I have unpacked the AV into three families of technologies: the electric engine, self-driving sensors, and shared ride algorithms. Let us now regroup them and consider how AV as a whole urges us to rethink our relations with transportation technologies: human intentionality no longer dominates the relations, as Ihde (1990) framed it in the original postphenomenological formula, but rather technological intentionality is becoming more active (Verbeek 2008a; Wellner 2017; 2018; 2020). I suggested a new model to accommodate an even more passive user to represent situations in which the AV decides which route to take, who are the passengers that will join the ride, and what music to play. These AV technologies do more than mediate. Their technological intentionality does not simply direct the passengers' behavior. They sometimes practically decide *instead* of their users.

In postphenomenology, most analyses focus on the human-technology micro-cosmos. Contemporary technologies like the AV demonstrate that this is not sufficient to examine how humans and things interact. There is a need to take the world into consideration. This becomes acute with AV technologies that exhibit stronger forms of technological intentionality. As technologies become "equipped" with a certain intentionality, the relations between humans and the world change. We are no longer "world builders" as Heidegger framed it (Uricchio 2011, 33). The world is no longer a picture. We are now "in" the world like parts of a huge machine controlled by algorithms, as Deleuze and Guattari forecasted.

Moreover, AV technologies do not just mediate a given world that is "out there," neutral and unchangeable. They change the world in various ways (cf. Aydin et al. 2019). The postphenomenological formula already includes the world, thereby reminding us to make it part of the analysis of human-technology relations. However, there is a need to further develop the role the world plays, beyond the micro-cosmos of I-technology (see Michelfelder 2015; Kranc 2017). More importantly, it may be an opportunity to raise awareness of the problem known as climate change and lead people to rethink the world in the context of the technologies they use and their responsibility for the changes in it.

In light of the Bruntland Report (1987) titled "Our Common Future," the reference to the world should consist of two elements: an endangered biosphere and a society composed of various human beings with diverse needs. The postphenomenological formula can be further developed to model these important elements.

Today, however, the discussion on environmental issues tends to focus mainly on reducing greenhouse gas emissions. Electric engines promise to reduce these emissions, although the emissions still exist in the everyday production of electricity, the production of batteries, and other occasions related to AVs. Batteries involve additional environmental considerations and so does the mere production of cars whether they are electrical or shared. The hope is that shared rides reduce emissions as people would share taxi rides, and in the long term may reduce the total number of private vehicles (cf. Greenblatt and Saxena 2015). The required focus on the world component can be answered from an architectural perspective seeking new city planning best practices, like repurposing of parking lots into green parks. Societal considerations can be answered by the algorithms if and when they ensure equal and safe access to shared rides, as well as the more remote issues of traffic control (see Mladenovic and McPherson 2016). These considerations can go back to the human element and seek to encourage walking as a measurement of promoting health.

The new technologies of AV can be an opportunity to make our world better. Our policies should ensure we promote shared rides rather than privately owned AVs, that these services will be accessible to all—physically, financially, and securely—and that the total mileage will be lower to ensure smaller emissions of greenhouse gases and pollution.

ACKNOWLEDGMENTS

The author thanks the Israeli Innovation Institute, the Nanooa project, https://www.nanooa.org.il, also known as the "Mobility Shift" Initiative, and its members for fruitful idea sharing on smart transportation and the future of cities in 2050.

NOTES

1. See https://www.theverge.com/2019/7/5/20683406/japan-car-sharing-renting-not-driving-private-space-orix-times24 (accessed July 8, 2019).
2. During the COVID-19 pandemic, this type of transportation gained traction in Tel Aviv because the number of passengers in the vehicle is much lower compared to buses. For some, however, this solution was not regarded as safe enough, and we have seen a surge in private car usages, at the expense of public transportation.
3. Jeffrey Greenblatt and Samveg Saxena found out that the combination of autonomous vehicle and shared transportation—that is, shared taxis—can significantly reduce greenhouse gas emissions in the United States (Greenblatt and Saxena 2015). The reduction is due to three factors: "battery-electric vehicles," smaller size vehicles, and higher annual mileage per vehicle. The authors tie together electric and shared vehicles, as if all shared rides must be performed by electric vehicles. They neglect the fact that autonomous driving as such has no effect on greenhouse gas emissions.
4. In order to collect passengers "on the fly" and take dynamic traffic conditions into account, the shared transportation vehicle (whether driven by a human driver or autonomously-automatically by an algorithm) must be networked—that is, connected to the internet.
5. The first age is equivalent to the age of the tool, with the difference that Deleuze and Guattari describe the relations as enslavement. The second age is equivalent to that of the machine and is described as subjection because the human operator is subjected to the technical machine (see G. Wellner 2016). They use the term machine in two meanings—one is the technical machine and the other is an apparatus, such as the state.

REFERENCES

Aydin, Ciano, Margoth González Woge, and Peter Paul Verbeek. 2019. "Technological Environmentality: Conceptualizing Technology as a Mediating Milieu." *Philosophy and Technology* 32 (2): 321–38.

Borenstein, Jason, Joseph R. Herkert, and Keith W. Miller. 2019. "Self-Driving Cars and Engineering Ethics: The Need for a System Level Analysis." *Science and Engineering Ethics* 25 (2): 383–98.

Deleuze, Gilles, and Felix Guattari. 1987. *A Thousand Plateaus: Capitalism and Schizophrenia*. Translated by Brian Massumi. Minneapolis and London: University of Minnesota Press.

Epting, Shane. 2018. "Automated Vehicles and Transportation Justice." *Philosophy & Technology* 1–15.

Greenblatt, Jeffrey B., and Samveg Saxena. 2015. "Autonomous Taxis Could Greatly Reduce Greenhouse-Gas Emissions of US Light-Duty Vehicles." *Nature Climate Change* 5 (9): 860–63.

Himmelreich, Johannes. 2018. "Never Mind the Trolley: The Ethics of Autonomous Vehicles in Mundane Situations." *Ethical Theory and Moral Practice* 21 (3): 669–84.

Ihde, Don. 1979. *Technics and Praxis: A Philosophy of Technology*. Dordrecht: Reidel Publishing Company.

Ihde, Don. 1990. *Technology and the Lifeworld: From Garden to Earth*. Bloomington and Indianapolis: Indiana University Press.

Irwin, Stacey O. 2014. "Technological Reciprocity with a Cell Phone." *Techné: Research in Philosophy and Technology* 18 (1/2): 10–19.

Kranc, Stanley. 2017. "Picturing the Technologized Background." *Kunstlicht* 38 (4): 20–27.

Latour, Bruno. 1992. "Where Are the Missing Masses? The Sociology of a Few Mundane Artifacts." In *Shaping Technology/Building Society: Studies in Sociotechnical Change*, by Wiebe E. Bijker and John Law, 225–58. Cambridge and London: The MIT Press.

Michelfelder, Diane. 2014. "Driving While Beagleated." *Techné: Research in Philosophy and Technology* 18 (1/2): 117–32.

Michelfelder, Diane. 2015. "Postphenomenology with an Eye to the Future." In *Postphenomenological Investigations: Essays on Human-Technology Relations*, edited by Robert Rosenberger and Peter-Paul Verbeek, 237–46. Lanham, MD: Lexington Books.

Mladenovic, Milos N., and Tristram McPherson. 2016. "Engineering Social Justice into Traffic Control for Self-Driving Vehicles?" *Science and Engineering Ethics* 22: 1131–49.

Santoni de Sio, Filippo. 2017. "Killing by Autonomous Vehicles and the Legal Doctrine of Necessity." *Ethical Theory and Moral Practice* 20 (2): 411–29.

Uricchio, William. 2011. "The Algorithmic Turn: Photosynth, Augmented Reality and the Changing Implications of the Image." *Visual Studies* 26 (1): 25–35.

Verbeek, Peter-Paul. 2008a. "Cyborg Intentionality: Rethinking the Phenomenology of Human–Technology Relations." *Phenomenology and Cognitive Science* 7: 387–95.

Verbeek, Peter-Paul. 2008b. "Morality in Design: Design Ethics and the Morality of Technological Artifacts." In *Philosophy and Design: From Engineering to Architecture*, by Peter Kroes, Pieter E. Vermaas, Andrew Light, and Steven A. Moore, 91–103. Dordrecht: Springer.

Wellner, Galit. 2014. "Multi-Attention and the Horcrux Logic: Justifications for Talking on the Cell Phone While Driving." *Techné Research in Philosophy and Technology* 18 (1/2): 48–73.

Wellner, Galit. 2016. *A Postphenomenological Inquiry of Cell Phones: Genealogies, Meanings, and Becoming.* Lanham, MD: Lexington Books.

Wellner, Galit. 2018. "From Cellphones to Machine Learning. A Shift in the Role of the User in Algorithmic Writing." In *Towards a Philosophy of Digital Media*, by Alberto Romele and Enrico Terrone, 205–24. Cham: Palgrave Macmillan.

Wellner, Galit. 2017. "I-Media-World: The Algorithmic Shift from Hermeneutic Relations to Writing Relations." In *Postphenomenology and Media: Essays on Human–Media–World Relations*, by Yoni Van den Eede, Stacey Irwin, and Galit Wellner, 207–28. Lanham, MD: Lexington Books.

Wellner, Galit. 2020. "Postphenomenology of Augmented Reality." In *Relating to Things: Design, Technology and the Artificial*, by Heather Wiltse, 173–87. London: Bloomsbury Visual Arts.

Wellner, Galit. 2022. "Becoming-Mobile: The Philosophy of Technology of Deleuze and Guattari." *Philosophy & Technology* 35 (2): 1–25.

Zuboff, Shoshana. 2019. *The Age of Surveillance Capitalism: The Fight for the Future at the New Frontier of Power*. New York: PublicAffairs.

Chapter 3

The Ethics of Crossing the Street

Robert Kirkman

CROSSING THE STREET

I am walking on the sidewalk alongside a busy avenue near my home in a small city. I am on the left-hand side of the avenue, in my direction of travel, and on the cross-street ahead a car is waiting to turn right. At least, that is what I perceive, as the right turn signal is blinking, and the driver is looking pointedly to her left to see if there is a gap in oncoming traffic.

She is still waiting there when I reach the intersection and must decide whether to step out into the crosswalk. I have the right of way, technically speaking, but that right would not amount to much if the driver were to seize her moment to turn just as I am crossing in front of her car. So I stop at the curb and wait, trying to make eye contact with the driver.

She looks briefly to her right and sees me. We exchange a glance, and with it a question and a response: I raise my eyebrows; she nods and waves me across. I smile, wave my thanks, and cross in front of her car.

This brief interaction between pedestrian and driver is an *ethical* encounter: matters of value are at stake, and either the driver, the pedestrian, or both may be called on to give an account of themselves, to take responsibility for their actions. Crossing the street may not be one of the great moral conundrums of the present age, like capital punishment or the emergence of gene-editing technology, but it is an instance of a kind of everyday, experiential ethics that makes up the fabric of human life in the world. That, at least, is a premise of this chapter.

One way to come at the ethical character of crossing the street is from the side of moral theory. A theory of duty or of right action would emphasize that the encounter between the pedestrian and the driver is governed by a set of moral principles rooted in the autonomy of each party as a moral being. The moral community of which they are part assents to a set of laws that confer rights and responsibilities on each party, and the assent of the community gives those laws binding force. So the pedestrian has the right of way and, in getting the attention of the driver, requests affirmation of that right. The driver ought to affirm the pedestrian's right and to act accordingly; to do otherwise would be morally wrong. Of course, it remains open to the two parties to negotiate a different outcome: the pedestrian may voluntarily yield the right of way, for example, whether as an act of kindness or in keeping with some overriding principle.

Utilitarian theory would emphasize instead that each of the possible outcomes of the encounter would involve benefits and harms for the driver, the pedestrian, and the wider community. The story as presented here may have the optimal outcome: the pedestrian crosses with relatively little risk of harm, the driver experiences only a momentary delay, and the averted harm and relative efficiency of the exchange will ramify out to the families and friends of each, to other drivers and pedestrians in the area, and to the wider community. Traffic laws that include a right of way for pedestrians also have risks and rewards for the community that, for a utilitarian, would be reason enough to enact and enforce them.

The power of theoretical frameworks lies in abstraction: they draw away from the messy details of the moment to provide a clear and consistent basis for judgment. They provide a view from the outside, from a neutral vantage point, and may best be used in hindsight. Looking back on the encounter at the intersection, with some time to think, I may judge that I did the right thing, that is, the thing that accords with one or another principle.

And yet, in the moment of the encounter, in the full richness of the present situation, it seems unlikely I would pause long enough for careful deliberation regarding principles. Instead I perceive and respond to the situation as it unfolds, establishing a fleeting but vital relationship with the driver who is also situated in the present moment. More than this, some of the messiness—or perhaps the richness—of the present moment may have bearing on how I may respond and act, details that would be lost in too quick a retreat into abstraction.

An experiential approach to ethics remains situated in place, in the present moment, revealing and describing the structure of the experience and the aspects of it that draw attention and motivate action. A phenomenological method is especially apt for a descriptive task of this sort.

Without going into too much detail, a phenomenology of ethical action brackets off or puts out of play moral theories, as well as moral psychology, cognitive theory, and other scientific accounts of human behavior, in order to bring to light the lived experience of action. To put theoretical frameworks out of play is not to abolish them or to deny their usefulness; it is simply a way to avoid taking them for granted or, as Merleau-Ponty (2012, 57) (Merleau-Ponty, 2012, p. 57) might put it, so that we remember their origins. The aim is to describe the primordial, pretheoretical values and motivations that give meaning to action in the world and so give meaning to the values and obligations articulated in moral theories.

I inhabit a world and pursue projects through it. The world I inhabit is an open field of horizons (Merleau-Ponty 2012, 345): what I perceive points beyond itself to what I do not perceive here, or do not yet perceive, or no longer perceive. I see a closed door from one side, but perception points beyond that to what is now on the other side of the closed door, or the possibility that the door may soon be opened. I may walk across the room and open the door, a project that gives meaning to my action and shifts the meaning of this room, this door, and whatever is on the other side of the door as I put the project into action (per Merleau-Ponty 2012, 115).

Value, in its most primordial sense, is the motive and the meaning of a project that points beyond a horizon; what rolls into view as I move and act in the world of horizons may surprise or disappoint, but, in any case, it engages my attention and draws action forward. A primordial value may not amount to a *reason* for acting, because it need never be expressed in propositional form; it may, in technical terms, be *pre-thetic*, not yet set down in some determinate formulation. A primordial value also need not be an *ethical* value, though ethical values too find the roots of their meaning in lived experience as soon as I encounter other people in the world.

As I walk toward the intersection, my actions are motivated and infused with meaning by an array of projects, including perhaps the overriding project of getting to the train station a few blocks up the avenue, but also delight in the musical pulse and flow of walking (Sacks 1987, 144–45). When I see the driver waiting at the intersection, I perceive another person who is "geared into" the world in the present moment, just as I am (Merleau-Ponty 2012, 367). I perceive her as acting with intention, pursuing projects of her own: she aims to turn safely into traffic, perhaps with the overarching aim of making it to an appointment on time. I hesitate before stepping off the curb, not because I have formulated reasons for my action or rendered a judgment on possible outcomes but because my own projects have come up against a complex set of future possibilities, including horizons I would rather not cross.

Describing my encounter with the driver in this way highlights a first step on the way from primordial values as such to primordial *ethical* values: my vulnerability in the world. If I were to step into the street just as the driver steps on the gas, the meaning and possibilities of the world of my lived experience might be forcibly rearranged, closing off any number of projects that matter or even, in an extreme turn of events, collapsing my world altogether and with it the hope of pursuing any projects at all. With such possibilities looming over the horizon, it is no wonder that I hesitate before stepping off the curb!

But I am not the only one who is vulnerable in the world. The driver is also vulnerable to the disruption of any or all of her projects, even if her vulnerability and mine in the present circumstances are not symmetrical; she is, after all, surrounded by and cushioned within a large machine. But suppose our situation were reversed, and I was the one behind the wheel, trying to turn right onto a busy avenue. A long-awaited gap in the traffic is about to open in front of me, but, just before I step on the gas, I glance to my right and see a pedestrian who intends to cross in front of my car. In that moment, I may become acutely aware of her vulnerability and, with that awareness may come the motivation to step more firmly on the brake pedal and focus on a nonverbal conversation with her.

In this reverse case may be found a second step toward ethical values: the possibility of perceiving the intentions of others, their projects, and the values toward which they move, as if they were in some sense my own. Nell Noddings (2013, 30) casts this mode of perception as *receptivity*:

> The notion of "feeling with" I have outlined does not involve projection but reception. I have called it "engrossment." I do not "put myself in the other's shoes," so to speak, by analyzing his reality as objective data and then asking, "How would I feel in such a situation?" On the contrary, I set aside my temptation to analyze and plan. I do not project; I receive the other into myself, and I see and feel with the other. I become a duality.

Note that receptivity is not a process of analysis based upon putatively objective facts but is, rather, a mode of perception through which I engage with the projects and the lived world of another person. As such, Noddings's account of receptivity meshes well with a phenomenological approach.

There remains one more step on the way to distinctly ethical values, because it remains possible for me to perceive the other's projects without those projects coming to matter to me. In the reverse situation, in which I am behind the wheel, I have any number of projects that draw my attention and solicit a response from me. If I am honest with myself, I would have to admit that many of them are focused on my own well-being and my own vulnerability: one aspect of my motivation for pressing on the brake and opening nonverbal negotiations with the pedestrian is to avoid legal liability and the other unpleasant (for me) possibilities that lurk behind the horizons that would open up if I were to strike her with my car. If those were my only motivations, I would be stuck in a kind of primordial callousness that does not even rise to the level of principled egoism.

Noddings (2013, 33) again provides a vital clue to this third step, in what she characterizes as a *motivational displacement*:

> When I care, when I receive the other in the way we have been discussing, there is more than feeling; there is also a motivational shift. My motive energy flows toward the other and perhaps, though not necessarily, toward his ends. I do not relinquish myself; I cannot excuse myself for what I do. But I allow my motive energy to be shared; I put it at the service of the other.

Primordial ethical values involve care for others and their projects—perhaps especially for their "basic needs and legitimate expectations," as Anthony Weston (2018, 7) phrases it—for their own sake.

The interaction between pedestrian and driver unfolds in the vivid present. It is not a case study calling for a judgment based on this or that principle but an immediate situation calling for engagement and response. How it all plays out, how the pedestrian and the driver each act, will come down to how they perceive one another and their shared surroundings, how their projects align with or accommodate one another, and whether they attend to and care for each other as vulnerable beings in the world. More fancifully, their interaction would seem to have much more in common with dancing than with litigating a court case.

AUTONOMOUS VEHICLES

Now suppose I am walking on the sidewalk alongside a busy avenue near my home in a small city. I am on the left-hand side of the avenue, in my direction of travel, and on the cross-street ahead a car is waiting to turn right. At least, that is what I perceive, as the right turn signal is blinking. The person sitting in the front seat, however, is focused intently on her tablet, reading or watching the news, or maybe chatting with a friend. The car has no steering wheel, which implies it is a model that promises Level 5 autonomy: the "driving automation system"—that is, the control system of an autonomous vehicle—performs the entirety of the "dynamic driving task," without any expectation that the human user will serve as a "fallback" by resuming direct control of the vehicle (SAE International 2016, 16).

As a matter of law, I have the right of way. Still, I stop at the curb and hesitate, unsure of the wisdom of crossing in front of an autonomous vehicle. This new situation differs markedly from the encounter at the intersection with a human driver, in at least four ways.

First, a driving automation system does not act with intention. In its taxonomy of driving automation systems, SAE International (2016, 5) analyzes the "dynamic driving task" (DDT) into its operational and tactical components, including "lateral vehicle motion control via steering (operational)" and "maneuver planning (tactical)," among others. It is a system performing calculations and causing changes in velocity and trajectory. A human driver, by contrast, is a living being who inhabits a world of horizons and acts with intention. In the act of driving, the mechanics of velocity and trajectory fade into the background: I do not have to calculate how many degrees to turn the wheel to execute a turn into a calculated trajectory but simply *turn right* because that is the way I am going. Merleau-Ponty (2012, 144) casts driving as an extension of motor intentionality:

> If I possess the habit of driving a car, then I enter into a lane and see that "I can pass" without comparing the width of the lane to that of the fender, just as I go through a door without comparing the width of the door to that of my body.

Driving in traffic is like dancing on a crowded floor, a matter of attuned responsiveness and flow, rather than a process of calculation in carrying out predefined operational and tactical tasks.

The second difference in the situation is a corollary of the first: the driving automation system does not perceive me as a person. Among the operational and tactical components of the dynamic driving task is "object and event detection and response," a subtask of the DDT that includes "monitoring the driving environment (detecting, recognizing, and classifying objects and events and preparing to respond as needed) and executing an appropriate response to such objects and events" (SAE International 2016, 12). In the encounter at the intersection, the driving automation system may detect me as an object and may have classified me as a pedestrian, and it may have calculated some probability as to how this putative pedestrian-object may move in the immediate future, but that is far from saying the system and I share a world in which we may encounter one another as persons.

This point emerges forcefully in a preliminary report from the National Transportation Safety Board (National Transportation Safety Board 2018) on a fatal accident involving an autonomous vehicle and a pedestrian in Tempe, Arizona, in March 2018.

> According to data obtained from the self-driving system, the system first registered radar and LIDAR observations of the pedestrian about 6 seconds before impact, when the vehicle was traveling at 43 mph. As the vehicle and pedestrian paths converged, the self-driving system software classified the pedestrian as an unknown object, as a vehicle, and then as a bicycle with varying expectations of

future travel path. At 1.3 seconds before impact, the self-driving system determined that an emergency braking maneuver was needed to mitigate a collision.

There are complications in the Tempe accident regarding the expected response of the human operator who was to serve as fallback, but the description of the software process is telling: at no point in its process of detection and classification did or could the system perceive a person.

The third difference in the situation is that the driving automation system does not have "skin in the game," which is to say, it is not *vulnerable*. In the encounter with a human driver, both driver and pedestrian are vulnerable in the world: each pursues projects that can fail utterly, including the most basic project of remaining alive. The mutual recognition of vulnerability is one of the bases for an ethical relationship between them, or at least the basis of a motivation to pay attention to one other. While a driving automation system can fail or even be destroyed, it is not aware of its own vulnerability. An autonomous vehicle cannot experience doubt or fear, it does not face any risk of losing its foothold in the world, if only because it does not inhabit a world; it merely exists in space.

The fourth difference, following from the other three, is that the autonomous driving system is not capable of motivational displacement: engagement with and care for others for their own sake. Programmers may have attempted to program ethical parameters into the system, but that alone does not establish the possibility of an ethical relationship between the car and people it may encounter. To base an "ethical algorithm" on utilitarian theory, for example, would be in effect to design the system to kill an optimal number of pedestrians, as there may be circumstances in which the vehicle would strike one or more pedestrians to avoid a greater harm, based simply on calculations of marginal net utility. Granted, the optimal number is *typically* zero, but, standing at the intersection, I have no way of guessing whether the calculations of the autonomous driving system will yield the result that this situation is typical.

In sum, what seems to be missing in my hypothetical encounter with an autonomous vehicle is the possibility of mutual acknowledgment, of negotiation based on the reciprocal recognition of shared vulnerability. I may manage to get the attention of the person in the front seat, someone capable of such acknowledgment and negotiation, but she may just shrug at me: the car will do what it is programmed to do. For all I know, the driving automation system may have detected my approach and factored it into the calculations for its next move, one way or the other, but it offers no external indication.

Suppose, though, that the driving automation system has been designed to be an "emotionally sentient agent," as some current research in artificial intelligence would have it. The promise of this research is not to create artificial persons who inhabit the world, engage in projects and feel emotions as humans do, but to "develop models of emotion that are amenable to computation" (McDuff and Czerwinski 2018, 76). What emotional sentience amounts to, then, is a system for detecting verbal and gestural cues and fixing them with emotional labels so that the machine can locate and execute responses a human user will perceive as appropriate. The aim is simply to make human users more comfortable with and more attached to their machines (McDuff and Czerwinski 2018, 74).

If the driving automation system before me is programmed as an "emotionally sentient agent," the most I could expect is for it to have a display that can mimic human facial expressions, or an external speaker that can emit an audio signal or synthesized speech. Even supposing it provided some signal to suggest it would be safe for me to proceed across the street, the ethical texture of the situation is essentially unchanged: the car has offered the mimicry of a human response directed to what the system registers as an object with a certain probability of being a pedestrian. Such mimicry would be entirely unlike a nod and a wave from a human driver as, again, the meaning of the latter springs from mutual recognition that may serve as a basis for trust.

So even if the driving automation system emits synthesized speech, even if it mimics a sympathetic gesture, the truth remains that I am alone at the intersection, left to make my own calculation of the probability of an impact between two bodies in space. The best I can do is to derive some small comfort from the legal and moral liability that might be incurred in case of a mishap by people who are absent—or simply disengaged—from the scene.

TECHNOLOGICAL FRAMES

The most common justification for a move to autonomous vehicles could well be called the argument from safety: the implementation of driving automation systems will reduce the risk of death or injury from automobile accidents. The basic premise seems reasonable enough: humans behind the wheel can be unreliable and, all too often, distracted, fatigued, or otherwise impaired. This fallibility of human drivers no doubt contributes to a rate of injury and fatality that, in other contexts, would be considered a public health crisis. Driving automation systems, by contrast, are supposed to be more reliable and, depending on the overall prevalence of autonomous vehicles and of the infrastructure to support them, may be able to coordinate with other vehicles to avoid collisions and other mishaps. On these premises, and assuming the new technology functions as advertised, it only makes sense to swap out an unstable and unreliable human control system for something demonstrably better.

The argument from safety assumes a linear relationship between the implementation of driving automation systems and the reduction of risk. This is to say that the number of human drivers might be regarded as an independent variable in a linear equation, the output of which is a rate of injuries and fatalities: an incremental reduction in the number of human drivers would yield an incremental reduction in risk. If it is indeed linear, this relationship holds across the range of possible values for the independent variable: reducing the number of human drivers to zero would not eliminate all risk, but it would reduce it to a minimum. Clearly, a lower risk of death is a good thing, and a minimal risk of death is even better, so who could be opposed to autonomous vehicles?

The argument from safety may be employed to poison the well, in effect, by implying that anyone who objects to the introduction of autonomous vehicles must be in favor of traffic deaths. Aside from that, the argument is flawed in that it understates the complexity of the situation. For one thing, the argument narrows the scope of ethical consideration to a single term, safety. Reduction of risk is surely important, all else being equal, but it is not a value to be secured regardless of the cost. There are tradeoffs involved—such as the loss of any possibility for reciprocal recognition between drivers and pedestrians—that might reasonably lead someone to conclude that it is worth accepting a degree of risk higher than what would be possible through a given intervention.

For another thing, the causal relationships among human fallibility, road safety, and other factors are not linear. This is simply to say that automotive transportation is a system, one that is nested within, shapes, and is shaped by the more encompassing metropolitan systems through which most people now inhabit landscapes. Systems involve complex interactions among components, with recursive feedback loops that, taken together, yield emergent behavior that is not linear. As Donella Meadows (2008, 91) observes,

> the world often surprises our linear-thinking minds. If we've learned that a small push produces a small response, we think that twice as big a push will produce twice as big a response. But in a nonlinear system, twice the push could produce one-sixth the response, or the response squared, or no response at all.

At a finer grain, automotive transportation constitutes a *technological* or *sociotechnological* system, a hybrid ensemble that includes technical components (hardware and software) and social components (institutions and values) (Hughes 1994, 104–6). Because they are hybrid ensembles, interactions within technological systems involve both material causality and exchanges of meaning: as a system develops, it may establish and reinforce the meaning of an artifact and expectations as to the role it should play in human life (see Pinch and Bijker 1987, 40–46).

As with systems in general, the behavior of technological systems is nonlinear: the introduction of an increment of technical change or institutional change may yield no change at all in the behavior of the system, or it may move the system over a tipping point into an abrupt and marked change. In any case, the unfolding of a technological system always exceeds the intentions and even the imagination of those whose decisions shape it.

It is now a matter of legend that, in its early days, the automobile was sometimes described as a "horseless carriage." The term frames the automobile as a simple, linear intervention in the system of transportation and urban planning: take a carriage, remove the horse, add an internal combustion engine, and the result will be *just like a carriage* but without all that manure and all those horse carcasses that need to be removed from the city. As it happened, the "horseless carriage" turned out to be *nothing at all* like a carriage. The automobile was an entirely different kind of thing. Not only did it interact differently with other elements in the inhabited landscape but, as its meaning stabilized, the automobile began to change the meaning of those other elements and, eventually, the shape and meaning of the inhabited landscape itself.

In other words, the automobile emerged as the boundary object of an entirely new technological frame, one that has come to dominate inhabited landscapes around the world. The term, "technological frame," was introduced by Weibe Bijker (1995, 122–24) to encompass "all elements that influence the interactions within relevant social groups and lead to the attribution of meanings to technical artifacts—and thus to constituting technology." The boundary object is the artifact around which and in response to which those interactions arise. For social actors not already included in the frame, "the artifact presents a 'take it or leave it' decision." To opt into the frame is to accept the artifact as it is along with the social interactions that have arisen with it; "if you buy a car, for example, you become included in the semiotic structure of automobiling: cars-roads-rules-traffic-jams-gas prices-taxes" (Bijker 1995, 284–85).

In the age of the carriage, it is not necessarily the case that there was a horse-drawn-carriage frame. The history of urban form suggests rather that the carriage fit within what might be called a walking-city frame (Jackson 1985, 14–20). Within that frame, the street was a multiuse public space, with carriages, pedestrians, cyclists, merchants, children, and livestock all sharing it according to rules and conventions that are now difficult to imagine.

When the automobile first appeared on the streets of cities and towns in England and in the United States, it would have been seen mainly as a noisy and dangerous plaything for adventurous young men with more money than sense. In his history of suburbanization in the United States, Kenneth Jackson (1985, 158) relates that many states and local governments in the United States followed the lead of England in passing laws placing strict limits on automobiles on city streets: they could travel at no more than four miles per hour and had to be preceded by a man carrying a red flag, presumably to warn the rightful users of the street of the approaching disruption.

How things have changed! Within the automotive frame, the street is now *primarily* an artery through which a stream of cars should be permitted to flow with the fewest possible impediments. Cyclists must wear bright clothing and mount flashing lights on their bicycles. Pedestrians must confine themselves to the sidewalk, if there is one, and cross the street with great care—if not trepidation—even where there is a legally sanctioned crosswalk.

Automobiles under the control of driving automation systems are sometimes described as "driverless cars": remove the driver from a car, add a computer and a bunch of sensors, and the result is supposed to be *just like a car* but without the flaky, distractible, sleepy, and generally fallible human being in control of it. To say it is "just like a car" is to say that substituting one control system for the other would leave the automotive frame, the structure of meanings with which the artifact is vested, unchanged.

If the analogy between a driverless car and a horseless carriage holds, though, inhabited landscapes may come to be enclosed within an entirely new technological frame, with driving automation systems as the new boundary object. By the nature of systems, it is difficult to foresee what life inside that frame would be like. After all, those whose decisions established the automotive frame foresaw only increased speed and convenience, not traffic jams, photochemical smog, geopolitical instability surrounding oil supplies, climate change, and pedestrian fatalities. Within a possible driving automation frame, the street may well become a conduit through which computers guide themselves on behalf of their hapless human passengers and, even more than at present, pedestrians, cyclists, and others must fend for themselves or stop using the street altogether.

Stepping back to the systems perspective provides a wider view of the encounter at the intersection: the negotiation between pedestrian and driver is framed within and conditioned by the whole ensemble of meanings, of opportunities and constraints, that arise within the shared social world with the establishment of the automobile as a boundary object. I may be just one pedestrian at one intersection, but the experience I have there and the values at play are part of the wider fabric of life in a metropolitan region. A change in the control system of the vehicle in front of me would transpose me into an entirely different ensemble of meanings, possible projects, opportunities, and constraints.

Chapter 3

STREETS WORTH CROSSING

These reflections on the act of crossing the street introduce an alternate perspective on the development and possible adoption of driving automation systems in the context of present-day metropolitan regions. Behind the question of whether to step off the curb at a given moment lie questions of the meaning of public spaces and of shared environments more broadly, which bear on the prospects of everyone who pursues projects within those shared environments. Responding to these questions is at least as much a matter for public deliberation and thoughtful policymaking as it may be for the reassurances of technical experts and the workings of the market.

Those who advocate for the widespread adoption of driving automation systems offer them up as a technical solution to a straightforward problem: fallible human drivers cause fatal accidents, so replacing them with purportedly less fallible control systems may reduce the number of fatal accidents. Removed from all context, the ethical case for autonomous vehicles seems straightforward: all else being equal, a reduction in the rate of death and injury from automobile accidents would be a good thing.

The view from the sidewalk suggests that all else is not equal, in part because the problems posed by automotive transportation are plural, nonlinear, and perhaps not amenable to anything so tidy as a solution, let alone a technical fix. As a corollary to his argument regarding problems that have no technical solutions, Garrett Hardin (1968, 1245) observes: "The morality of an act is a function of the state of the system at the time it is performed." So any deliberation regarding driving automation systems should be attentive to its systemic context.

The current state of the system has come in for its share of criticism. Streets work well enough for some people engaged in some projects but also throw up obstacles in the way of others. Drivers may be stuck in traffic, pedestrians unable to cross safely, and all may find themselves beset by confusing signals, stress, alienation, and even rage. The question for deliberation is how the whole system might be transformed by the widespread adoption of autonomous vehicles, and how the horizons of possibility shift for people pursuing their various projects.

The standard argument for autonomous vehicles also suffers from too narrow a focus on numbers as proxies of value. Yes, fewer deaths would be better, all else being equal. But people pursue many projects, aimed at different kinds of values, not all of which can be measured and compared with one another. This is not to say utilitarian values, which do lend themselves to expression as numbers, are not important: it matters whether an action makes people better or worse off in ways that may be measured and compared. It also matters whether people act out of respect for the autonomy of others, or out of commitment to care for others, or from well-calibrated habits of character. Beyond these may also be found more fine-grained values of experience and relationship and purpose that give shape and meaning to human life in the world, all of which matter even if it is not always easy to put them into words. Taking all such values into account, a city dominated by autonomous vehicles might well be incrementally safer, in the aggregate, but might also be a landscape of alienation and dread for anyone not currently strapped into a passenger seat.

Attending to systems and to the values to be discovered in lived experience makes possible a radical reframing of the problems at hand. Even if the goal of reducing the number of automobile fatalities is of paramount importance, perhaps a more radical innovation to that end would be to break out of the automotive frame, to consider how streets and cities might be reconfigured so that people can access what they need and what they want without cars of any kind, driven or driverless. Policies and practices to that end would be many, difficult to implement, and uncertain in their results, but they would at least constitute a systemic response to a systemic problem, one that may be attentive to the kinds of fine-grained values that, among other things, would make a street worth crossing.

REFERENCES

Bijker, Wiebe E. 1995. *Of Bicycles, Bakelites, and Bulbs: Toward a Theory of Sociotechnical Change.* Cambridge, MA: MIT Press.
Hardin, Garrett. 1968. "The Tragedy of the Commons." *Science* 162 (3859): 1243–48.
Hughes, Thomas P. 1994. "Technological Momentum." In *Does Technology Drive History?: The Dilemma of Technological Determinism*, edited by Merritt Roe Smith and Leo Marx, 101–13. Cambridge, MA: MIT Press.
Jackson, Kenneth T. 1985. *Crabgrass Frontier: The Suburbanization of the United States.* Oxford: Oxford University Press.
McDuff, Daniel, and Mary Czerwinski. 2018. "Designing Emotionally Sentient Agents." *Communications of the ACM* 61 (12): 74–83.
Meadows, Donella H. 2008. *Thinking in Systems: A Primer.* Edited by Diana Wright. White River Junction, VT: Chelsea Green Pub.
Merleau-Ponty, Maurice. 2012. *Phenomenology of Perception.* Translated by Donald A. Landes. New York: Routledge. Doi:40020498775.
National Transportation Safety Board. 2018. *Preliminary Report: Highway: Hwy18mh010.* Washington, DC: National Transportation Safety Board. https://www.ntsb.gov/investigations/AccidentReports/Reports/HWY18MH010-prelim.pdf.

Noddings, Nell. 2013. *Caring: A Relational Approach to Ethics and Moral Education.* Second, updated edition. Berkeley, CA: University of California Press.

Pinch, Trevor J., and Wiebe E. Bijker. 1987. "The Social Construction of Facts and Artifacts: Or How the Sociology of Science and the Sociology of Technology Might Benefit One Another." In *The Social Construction of Technological Systems: New Directions in the Sociology and History of Technology*, edited by Wiebe E. Bijker, Thomas P. Hughes, and Trevor J. Pinch, 17–50. Cambridge, MA: MIT Press.

Sacks, Oliver. 1987. *A Leg to Stand On.* New York: Harper Perennial.

SAE International. 2016. *Taxonomy and Definitions for Terms Related to Driving Automation Systems for On-Road Motor Vehicles.* SAE International. https://doi.org/10.4271/J3016_201609.

Weston, Anthony. 2018. *A 21st Century Ethical Toolbox.* Fourth edition. New York: Oxford University Press.

Chapter 4

Who Is Responsible If the Car Itself Is Driving?

Sven Ove Hansson

Self-driving cars have been discussed and promoted at least since the 1950s, and safety has always been a major argument for their introduction (see figure 4.1). However, even if they can be made safer than conventional vehicles, few would deny that accidents will still occur. Who will then be responsible? In conventionally driven cars, the driver is responsible for avoiding accidents and for the safety both of the passengers and of others who travel or walk on the same roads. One of the major issues that have to be solved before self-driving cars are introduced is who—if anyone—will take over the responsibilities previously carried by the driver.

In order to analyze the responsibility issues involved in this technology, we need to distinguish between different types of responsibility and clarify the quite intricate connections between our ascriptions of responsibility, causality, and blame.

WHAT IS RESPONSIBILITY?

The best starting point for discussing the meaning of "responsibility" is still the classification of responsibility concepts introduced by the British legal philosopher H. L. A Hart (1907–1992). Hart distinguished between four meanings of the term "responsibility."

- Hart's *liability-responsibility* can, in a moral context, be expressed as "deserving blame" or being "morally bound to make amends or pay

Figure 4.1. Experiments with self-driving cars, guided by electromagnetic signals from cables under the road surface, began in the 1950s. In January 1956, this picture appeared in an advertisement in Life Magazine with the caption: "One day your car may speed along an electric super-highway, its speed and steering automatically controlled by electronic devices embedded in the road. Travel will be more enjoyable. Highways will be made safe—by electricity! No traffic jams . . . no collisions . . . no driver fatigue."

compensation." In a legal context, it refers to being "liable to punishment" or "liable to be made to pay compensation" (Hart 2008, 225).
- By *role-responsibility*, he meant the "specific duties" a person holds through occupying "a distinctive place or office in a social organization" (2008, 212). His usage of "role" covers not only professional and official functions but also private roles such as those of spouse, parent, or host.
- With *causal-responsibility* he referred to cases in which the phrase "was responsible for" can be replaced by "caused" or "produced" without a change in meaning. According to Hart, causal responsibility can be attributed not only to human beings "but also to their actions or omissions, and things, conditions, and events." One of his examples

was: "The icy condition of the road was responsible for the accident" (2008, 214).
- By *capacity-responsibility* he meant that one has a sufficient degree of capacity to understand, reason, and control one's own actions. This is what we refer to with the phrase "he is responsible for his actions" (2008, 227).

Hart noted that all four of these meanings appear in both legal and moral uses of the term "responsibility" (2008, 215). There are close connections between legal and moral norms, and some legal theorists maintain that legal responsibility should closely follow moral responsibility.[1] Here the main focus will be on moral, rather than legal, responsibility.

Concerning *liability-responsibility*, Hart himself noted that the term "blame" would be more suitable in a moral context than "liability." Several authors, including Goodin (1987, 167), use the term "blame responsibility" instead. Goodin's terminology will be used here. It refers generically to the negative consequences that a person morally deserves as a result of doing something wrong. This includes both blameworthiness and the moral requirement to make amends for what one has done.

Hart's notion of *role-responsibility* refers to what one has to do or achieve. Several authors have noted that the term "role" tends to obscure the generality of this notion (Cane 2002, 32). Suppose that while you are waiting for a train at the railway station, a stranger asks you to watch her suitcase while she buys a cup of coffee. If you agree, you undertake a certain responsibility. It seems to fall under Hart's category of role-responsibility, but it is doubtful whether there is a social role associated with it. Some authors, for instance Gerald Dworkin, have retained Hart's terminology but interpreted the term "role" very broadly (Dworkin 1981, 29). Others, such as Robert Goodin, prefer the more encompassing term "task responsibility," which refers to the assignment of "duties, jobs or (generically) tasks" (Goodin 1987, 168). This should also include undertakings and agreements, which, as Peter Cane (2002, 32) points out, are also important sources of this type of responsibility. In what follows, I will use the term "task responsibility" for the responsibility to do or achieve something.

The way Hart uses the term *causal-responsibility* makes it difficult to see why he did not instead use the term "causality" for this notion. For instance, an icy road can certainly be the *cause* of an accident, and someone who should have sanded it can then be responsible for the accident, but what is gained by saying that the road itself is responsible? It is better to use the term "causality" for inanimate objects.

There are two ways in which a person can cause something. First, one can do so in the same sense as an inanimate object. If I lie asleep on the floor of a dark room, and you stumble upon me when entering the room, then I cause your fall in the same way a stone could have done. But if I am awake and stretch out my leg when you pass by, then my causal role is different. It is a role that only an agent can have, and "his, her or its agency serves to explain" the pertinent outcomes, which "can therefore plausibly be treated as part of the agency's impact on the world" (Honoré and Gardner 2010). It is not unreasonable to use the term "causal responsibility" for this notion, but here it will be called "agent causality."

As Hart pointed out, most adults are considered to have *capacity-responsibility*, but it is "lacking where there is mental disorder or immaturity" (Hart 2008, 218). We can speak of a person as responsible for her actions, in this sense, even if we do not know of any particular action that she is responsible for. Therefore it is better to treat capacity-responsibility as an *ability* to be responsible than as a form of responsibility per se. It is closely related to the notion in medical ethics of "capacity for autonomous choice" (often also called "decision-making capacity" or "competence"), which marks the distinction between persons who can or cannot give informed consent to a medical intervention. In what follows, the term "capacity" will be used for this notion.

This leaves us with two basic forms of (moral) responsibility, namely, blame responsibility and task responsibility, along with two responsibility-related concepts for which we use other terms, namely, agent causality (Hart's causal-responsibility) and capacity to be responsible (Hart's capacity-responsibility). It is common in the literature to focus on blame and task responsibility, but they are often denoted with terms that indicate temporal relationships. Blame responsibility has been called "backwards-looking responsibility" (van de Poel 2011), "retrospective responsibility" (Duff 1998), and "historic responsibility" (Cane 2002, 31). Task responsibility has been referred to as "forwards-looking responsibility" (van de Poel 2011) and "prospective responsibility" (Duff 1998; Cane 2002, 31). The temporal terminology is somewhat misleading. We can refer in retrospect ("historically") to someone's task responsibility. We can also consider prospectively whether our actions will in the future give rise to blame responsibility (Hansson 2007). The temporal terminology will not be used here.

CAUSES AND CAUSAL FACTORS

Both blame and task responsibility are closely connected with causality, and more specifically with agent causality. Concerning blame responsibility, you are for the most part "only morally responsible for what you cause" (Bernstein 2017, 165). Similarly, it would seem inappropriate to assign task responsibility to someone for something that she cannot, as an agent, bring about. These connections with agent causality have a crucial role in connecting blame and task responsibility with each other.

In many practical situations, these two major forms of responsibility come together. Suppose that your neighbor holds a noisy party that keeps you awake all night. She is then blame responsible for your sleepless night, which she caused. She is also task responsible for ensuring that this does not happen again. But in many other cases, the two types of responsibility do not coincide. Suppose that a motorist drives drunk and runs over a pedestrian. She is then blame responsible for the accident, and she is also task responsible for not repeating the mistake of driving inebriated. However, in order to prevent similar accidents from happening again, measures are needed that affect all drivers, not only this particular driver. The task responsibility for such measures must fall to public authorities and others who can influence drivers in general. Consequently, blame responsibility and task responsibility will go apart.

In everyday reasoning, we usually apply a simple one-cause model (or thought pattern) for causal attributions. When searching for the causal background of an event, we have a strong tendency to pick out one previous event, which we call "the cause." However, as was shown already by John Stuart Mill ([1843] 1996, 327–34), most events have a complex causality, and they are best described as the outcome of a whole set of causal factors. When we single out one of these causal factors as "the cause" of the event, we perform a gross simplification (Hoover 1990). It can be likened to turning off all the lights on a mass scene on a theater stage, with the exception of a single spotlight that puts all the light on only one actor. This can be done in many ways, each of which represents a particular perspective on the drama that is developing on the stage.

Suppose that a police patrol decided to chase a car whose driver did not stop when ordered to do so. The pursued driver entered an unlit road, where he crashed into an illegally parked truck and was killed. What was the cause of his death? We can easily identify four causal factors, namely, the driver's decision not to follow the orders of the police, the decision by the police to pursue the car, the illegal parking of the truck, and the lack of road lighting. Presumably, the accident would not have happened if any of these four factors were missing. Each of them can be designated "the cause" in some of the many perspectives that can be applied to this accident. For instance, a critic of police pursuits would describe the actions of the police as "the cause of the accident," whereas a lawyer defending the police officers would say that it was caused by the deceased driver's disobedience to police orders. They both select one of the causal factors to be "the cause" and treat the others as subsidiary background conditions.

The process of selecting one of several causal factors as "the cause" can be called *causal singularization*. Its outcome will depend on what perspective we apply. We choose the causal factor that seems, in the chosen perspective, to be the most important, most salient, and/or most influenceable. For instance, in a course on microbiology we learn that cholera is caused by *Vibrio cholerae*. In a course on public health we learn that it is caused by lack of proper sanitation (Rizzi and Pedersen 1992). These answers do not represent different medical opinions. Instead they are divergent emphases that are adequate for different purposes. A physician treating cholera patients has reasons to focus on the bacterium, whereas someone working with disease prevention had better focus on the eliminable environmental factors that contribute to the disease.

Making a long story short (and somewhat simplified), we can describe the identification of causal factors as a mainly scientific or science-based undertaking, whereas causal singularizations are largely based on social conventions. For instance, there is a longstanding tradition in medicine to put focus on those causal factors that are plausible targets of medical interventions. In the case of cholera, this convention is operationalized differently by physicians specializing in infectious diseases and in public health because their focus is on different types of interventions.

THE ETHICS OF CAUSAL SINGULARIZATION

Ideally, we want our moral deliberations to be based on factual appraisals that are untainted by our moral values. Consequently, our (moral) deliberations on responsibilities would have to be based on ascriptions of causality that are independent of our moral values. But unfortunately that does not seem to be possible. The reason for this is that in the process of causal singularization, we are predisposed to assign the role of "the cause" of a negative outcome to actions or activities of which we morally disapprove, whereas we avoid assigning it to actions or activities that we consider to be morally unassailable. Let us consider a few examples:

> *Example 1:* Due to massive rainfalls, a segment of the riverbank has been undermined, and anyone entering the area runs the risk of being drawn into the dangerous rapids.
>
>> **Case i**: Charles is well aware that a large part of the foundations of the riverbank has been swept away. In spite of this, he recommends Andrew to go all the way down to the river to look for fish. The bank collapses, and Andrew drowns in the rapids.

Case ii: Charles has no means of knowing that the riverbank is damaged. He recommends Andrew to go all the way down to the river to look for fish. The bank collapses, and Andrew drowns in the rapids.

In the first case, it seems reasonable to claim that Charles's ill-considered advice was "the cause" of Andrew's death. We would probably not hesitate to say that Charles "caused" the accident. In the second case, these statements seem much less reasonable. Although Charles's advice is a causal factor in both cases (presumably, the accident would not have happened without it), we are unwilling to call it "the cause" of the accident in the second case. A plausible explanation of the difference is that if we say that Charles's advice caused the accident, then this will implicate that Charles was (blame) responsible for the accident.

Example 2: Despite her parents' advice to the contrary, Anne goes for a long walk on a very cold winter day, wearing only thin summer clothes and no coat or jacket. Three hours later she calls her parents from a hospital, where she is treated for severely frostbitten toes. "It's so unfair," she sobs. "Why should this happen to me of all the thousands of people who were out there in the streets?"

"But dear Anne," says her mother. "I am sure they all had much warmer clothes than you. In this weather it is almost certain that you will have a cold injury if you dress like you did. No doubt, your way of dressing was the cause of your injury."

The mother's causal conclusion seems justified enough. But let us also consider what happened in the following summer:

Example 3: Despite her parents' advice to the contrary, Anne goes for a long walk in the late summer evening, wearing an unusually skimpy dress. Three hours later she enters a police station, weeping incessantly, to report a rape.

In the trial three months later, the defendant's attorney says, "There were thousands of women out in the streets that evening. In all probability, Anne was the only one who wore such an unusually revealing dress. We have just heard my client telling us that this is what made him approach her—admittedly in a somewhat pushy manner—rather than someone else. Given what we know about young men in this city I am convinced that if he had not approached her in this way, then someone else would have done so. It is therefore obvious that her dress was the dominant causal factor that led up to the interactions that we are here to clarify. I do not hesitate to say that her way of dressing was *the cause* of what happened."

In spite of the structural similarities with Example 2, we are not willing to accept the attorney's causal conclusion. The reason for the difference is of course that if we call Anne's behavior "the cause" of the rape, then that would take us a long way toward assigning blame responsibility to her, which would be morally absurd.

> *Example 4:* A man walks out into the street on a pedestrian crossing. The driver of an approaching car tries but fails to stop, and the man is killed.

> *Example 5:* A man steps out into a motorway where no pedestrian access is allowed. The driver of an approaching car tries but fails to stop, and the man is killed.

In Example 4, we are much less willing than in Example 5 to describe the pedestrian's own behavior as "the cause" of the accident.

These examples show that agent causality has strong moral implications. As Ludwig Wittgenstein said, "Calling something 'the cause' is like pointing and saying: 'He's to blame!'" (Wittgenstein 1976, 410). This is corroborated by psychological research showing that we are more prone to call an action "the cause" of a subsequent event if it violates some norm that we endorse (Lagnado and Gerstenberg 2017, 584–85). We may of course wish to purge our everyday notion of causality of its moral connotations. However, these connotations are deeply entrenched in our linguistic habits and ways of thinking, and it is highly doubtful whether such a purge has a reasonable chance of success.

BACKGROUNDING

The strong connection between causality and (both blame and task) responsibility has important political and social consequences: public perceptions of causality have a strong role in determining which responsibilities are allocated and to whom. In particular, if you manage to make the public conceive a particular causal factor as part of the causal background, then it will mostly be left out of discussions about responsibility and about policy in general. We can use the term *backgrounding* for a process leading to a causal factor being treated as a background factor rather than highlighted as a cause.

Organizations that contribute to health risks and other dangers are particularly active in backgrounding. Tobacco companies are a prime example. They produce, sell, and promote a product that causes premature death in about half of its habitual consumers (Boyle 1997, 2). Most of them became addicted before reaching the age of majority. How is it possible for these companies to disclaim responsibility for all these deaths? The answer is that they do it by backgrounding, that is, by claiming that "the cause" that a person smokes is a free and voluntary choice by herself.

For another illustration of backgrounding, consider the following two examples. The first is hypothetical:

Example 6: A manufacturer of chainsaws sells a model with a very powerful motor. The user regulates the speed of the chain by pressing a handle. If the handle is pressed to the bottom, then the chain will move so fast that the user cannot control the saw, and there are grave risks both to the user her- or himself and to people in the vicinity. The saw has an instrument from which users can see if they press the handle too hard, and it is legally prohibited to pass certain marks on that instrument. But in spite of this, accidents are common, and hundreds of people die every year due to chainsaws being run at too high speeds.

What caused the chainsaw accidents? I have as yet never encountered a person who said that these accidents are caused by careless users of the saw. Instead we tend to causally attribute the accidents to the dangerous construction of the saw. This causal view supports the standpoint that the manufacturer is responsible for the accidents and should therefore urgently provide the saws with a speed limiter that prevents them from being run at too high speeds.

Our next example is quite similar, but it is not hypothetical:

Example 7: A manufacturer of motor vehicles sells a model with a very powerful motor. The user regulates the speed of the vehicle by pressing a pedal. If the pedal is pressed to the floor, then the vehicle will move so fast that the user cannot control it, and there are grave risks both to the user her- or himself and to people in the vicinity. The vehicle has an instrument from which users can see if they press the pedal too hard, and it is legally prohibited to pass certain marks on that instrument. But in spite of this, accidents are common, and hundreds of thousands of people die every year due to motor vehicles being run at too high speeds.

In this case, we tend to consider the accidents to be caused by the users (drivers), and consequently, the consumers rather than the manufacturer are held responsible for the accidents. Therefore, as noted by Christer Hydén, "the most obvious measure to treat non-compliance of speed rules—the vehicle speed limiter—is not on the agenda yet" (Hydén 2019, 4). Estimates based on experiments with speed limiting devices indicate that obligatory speed limiters have the potential to reduce road fatalities by about 25 to 50 percent (2019, 5).

Similarly, most motor vehicles lack alcohol interlocks, and they can therefore be driven by an inebriated driver. However, drunk driving is forbidden, so why are cars allowed that can be driven by an inebriated person? In this case as well, the construction of the vehicles has been backgrounded, and the responsibility is left to the individual drivers (Grill and Nihlén Fahlquist 2012).

FOREGROUNDING

Backgrounding is not irreversible. Causal factors that linger in the background can be brought to attention and included in discussions on responsibilities and social change. We can call such a process *foregrounding*. Not uncommonly, social criticism puts a focus on causal factors that were previously treated as unalterable parts of the social structure. One important example of foregrounding was achieved by the movement for workplace health and safety that emerged in the late nineteenth century, with labor unions as the main driving force. Previously, dangerous working conditions were treated as unavoidable, and workplace accidents were blamed on the victims. Today it is generally accepted that workplace accidents are caused by dangerous working conditions, which employers are responsible for eliminating.[2] Currently, foregrounding is an important part of efforts to tackle major health problems such as smoking, malnutrition, and obesity. They cannot be solved unless we look beyond the choices of affected individuals and address the social conditions under which these so-called lifestyle choices are made. The same applies to the ongoing tragedy of traffic fatalities—the current death toll is about 1.35 million per year (WHO 2018).

An important feature of foregrounding is that it tends to put focus on causal factors that can efficiently be abated, even if these are factors for which no one is currently held (blame) responsible. Often this has the effect of disassociating blame and task responsibilities from each other and putting increased emphasis on the latter. For instance, if we want to put an end to youth criminality in a residential area, it is not sufficient to focus on the young people there who are (blame) responsible for the crimes. We have to put emphasis on preventive measures that can be taken by the local schools, social services, law enforcement agencies, voluntary organizations, etc. This means that we assign task responsibilities to people and organizations that (contrary to the delinquent youth) do not have much blame responsibility in the matter.

When "new" responsibilities are assigned as a result of foregrounding, this does not necessarily reduce any of the preexisting responsibilities. For instance, suppose that a school in a disadvantaged area takes responsibility (and obtains resources) for providing adequate individual help to all students who need it in order to fulfil the learning goals. In no way does this reduce the responsibility of each individual student for her own studies. To the contrary, the purpose is to facilitate her fulfilment of these responsibilities. Similarly, if road authorities assume task responsibility for reducing risks at pedestrian crossings, for instance by introducing speed bumps, then this does not reduce the task responsibility of drivers to drive safely at these crossings. Neither has the blame responsibility of a failing driver diminished. More generally speaking, neither blame nor task responsibility is distributed in a zero-sum fashion.[3]

In road traffic safety, there has been a considerable movement in the past two decades toward the foregrounding of previously neglected causal factors. This is largely due to Vision Zero, which was first adopted by the Swedish Parliament in 1997 and has since then been adopted (with some variations) in a large number of countries, as well as in many states and major cities in the United States (Rosencrantz et al. 2007; Belin et al. 2012). The basic assumption of Vision Zero is that "from an ethical standpoint, it is not acceptable that any people die or are seriously injured when utilizing the road transportation system" (Government Bill 1996/97:137, 15). Consequently, all serious accidents are treated as unacceptable, and efforts to reduce the number of fatalities and serious injuries must continue assiduously as long as accidents still occur. This cannot be achieved with the traditional approach that assigns almost the whole burden of responsibility to drivers and other road users. Therefore Vision Zero makes the designers and implementers of the transport system responsible for eliminating human deaths and injuries. The movement for Vision Zero is an unusually clear example of a movement for the foregrounding of previously backgrounded causal factors.

SELF-DRIVING CARS

We now have the concepts we need to discuss who will, in a self-driving car, have the responsibility for safety that the driver has in a conventional car. Four options are available.

- We can assign this responsibility to the car itself, or more precisely to the *artificial intelligence* built into it.
- We can hold the *users* who travel in the car responsible.
- We can treat traffic accidents in the same way as natural accidents such as tsunamis and strokes of lightning, for which *no one* is held responsible.
- We can hold *other persons* than the users responsible.

According to Mladenovic and McPherson (2016, 1135), "the core goal of self-driving vehicle technology is to allow the vehicle to take over responsibility for real-time driving decisions." However, it is highly doubtful whether this is true if we take the word "responsible" to have the same meaning as when we assign (blame or task) responsibility to a human being. In this discussion, it is important to distinguish between the particular issue of self-driving cars and the more general question of whether some future artificial agents will be assigned blame or task responsibility in the same way as human agents (Nyholm 2018a, 1209–10). As far as we know, this may very well ensue, but the computer systems running self-driving vehicles will probably not be plausible candidates for such treatment. In the discussion on artificial intelligence, several features have been proposed as crucial for us to perceive a robot as a peer to whom we will relate as to another moral agent. Prominent on these lists are beliefs, desires, emotions, capacity for self-reflection, and ability to react adequately to human emotions (Kroes and Verbeer 2014). The systems governing self-driving cars will be constructed to execute orders given by humans, and at least in the foreseeable future they cannot be expected to have any of these features.

If a vehicle is fully automated so that the user can sleep when traveling, it would seem blatantly unfair to assign responsibility for its performance to the user. This option can therefore be written off for fully automated vehicles (but not for semiautomated vehicles in which the user has to stay prepared to intervene when needed). Alexander Hevelke and Julian Nida-Rümelin have proposed a form of collective (blame) responsibility, shared by all users of fully automated vehicles. Such as scheme "might work like a tax or a mandatory insurance, possibly partly based on the number of miles driven per year" (Hevelke and Nida-Rümelin 2015, 626–27). However, this would only cover one aspect of blame responsibility, namely, compensation. It could not include criminal charges, another major aspect of blame responsibility. Even more importantly, it is difficult to see how task responsibility could be attributed to the collective of drivers, which is not in control of major targets of improvement, such as the construction of cars, roads, and communication systems.

In a much-discussed paper on military robots, Robert Sparrow warned against a situation in which no one can be held (blame) responsible "when an autonomous weapon system is involved in an atrocity of the sort that would normally be described as a war crime" (Sparrow 2007, 62). In subsequent discussions, such a situation has been called a "responsibility gap" (Champagne and Tonkens 2015) or "retribution gap" (Danaher 2016). Responsibility gaps may become a problem for future automated systems that are so advanced and human-like that we do not know whether or not we can assign responsibility to them. It seems to be a less probable outcome for less advanced automatons. Suppose that a Metro company installs advanced automated gates that are programmed to recognize travelers who bought tickets in an equally advanced ticket machine. One day when you enter a Metro station, the gate runs amok and beats your knees so hard that you cannot walk for several weeks. When you call the Metro company to complain, they tell you that the gate is fully automated, and therefore no one—and most certainly not the company—is responsible for what it does. You would presumably not accept that answer. Even if a machine is automated, someone must be responsible for it—be it the owner, the user, or the manufacturer. Exactly the same argument can be applied to an automated vehicle. It is difficult to see how we could accept a responsibility gap in terms of either blame or task responsibility for any vehicle constructed and used by humans. We are therefore left with the fourth alternative listed previously, namely, to assign (blame and task) responsibility for self-driving cars to other humans than those who travel in them.

By far the most plausible recipients of the responsibilities no longer carried by vehicle users are the "system owners," that is, those responsible for the construction of the vehicles and for the roads and communication systems used by the vehicles. A transfer of responsibilities to them will be a continuation of the previously mentioned transfer of responsibilities in connection with Vision Zero. Furthermore, the early experience of self-driving cars points in the same direction. After the first serious accidents, the responsibility of the car manufacturers was emphasized. This is also what the automobile industry appears to be planning for (Atiyeh 2015; Nyholm 2018b).

However, this is not the end of the story. When several large organizations share responsibilities, we cannot expect their sharing to be free of conflicts. There will be choices between safety measures on infrastructure and on vehicles, for instance between improved detectability of lanes and improved lane detection systems in vehicles. Because infrastructure is usually paid by public means and vehicle improvements by the private sector, such choices can easily lead to political conflicts on the division of costs.

We can also expect cost distribution disputes within the private sector. The automobile industry has a history of protracted blame games. The most well known of these emerged in the wake of a series of severe accidents with Firestone tires mounted by Ford on their Explorer model. Countermeasures were delayed by years of mutual blaming. Initially, Firestone also blamed the drivers and owners of the vehicles. After considerable procrastination, Ford finally issued a voluntary recall of 6.5 million tires in August 2000 (Noggle and Palmer 2005). Unfortunately, there are signs that blame games targeting the owners and users of automated vehicles are far from implausible.

> Another, more tragic incident in May of 2016 also marked a first in the history of robotic driving. A Tesla model S in "autopilot" mode collided with a truck that the Tesla's sensors had not spotted, leading to the first fatal crash: the man riding in the Tesla vehicle was instantly killed when the truck and the car collided. . . . In a carefully worded statement, Tesla expressed sympathy for their customer's family; but Tesla also emphasized that the "customer" is "always in control and responsible." (Nyholm 2018a, 1202)

Summing this up, the most plausible scenario is that for driverless vehicles, the responsibilities carried by the drivers of conventional cars will be transferred to the constructors and maintainers of the vehicles, the road infrastructure, and the communication systems that guide and coordinate the vehicles. However, for this to work in practice it is necessary that legal rules and instruments are created to ensure that these organizations do not shift over responsibilities to road and vehicle users or engage in infighting that delays their response to urgent safety concerns.

NOTES

1. For instance, Michael Moore maintains that "legal liability in (torts and criminal law) falls only on those who are morally responsible" (Moore 2009, vii).

2. Attempts are still made to blame the victims and thereby relieve employers of responsibility for the working conditions. For an example of this argumentation, see Machan (1987), and for a rebuttal, see Spurgin (2006).

3. Concerning blame responsibility, this has been clearly shown in the legal literature. See, for instance, Moore (1999, 9–13) and Bernstein (2017).

REFERENCES

Atiyeh, Clifford. 2015. "Volvo Will Take Responsibility of Its Self-Driving Cars Crash." *Car and Driver*, October 8. https://www.caranddriver.com/news/a15352720/volvo-will-take-responsibility-if-its-self-driving-cars-crash.

Belin, Matts-Åke, Per Tillgren, and Evert Vedung. 2012. "Vision Zero—A Road Safety Policy Innovation." *International Journal of Injury Control and Safety Promotion* 19 (2): 171–79.

Bernstein, Sara. 2017. "Causal Proportions and Moral Responsibility." In *Oxford Studies in Agency and Responsibility*, volume 4, edited by David Shoemaker, 165–82. New York: Oxford University Press.

Boyle, Peter. 1997. "Cancer, Cigarette Smoking and Premature Death in Europe: A Review Including the Recommendations of European Cancer Experts Consensus Meeting, Helsinki, October 1996." *Lung Cancer* 17 (1): 1–60.

Cane, Peter. 2002. *Responsibility in Law and Morality*. Oxford: Hart Publishing.

Champagne, Marc, and Ryan Tonkens. 2015. "Bridging the Responsibility Gap in Automated Warfare." *Philosophy and Technology* 28 (1): 125–37.

Danaher, J. 2016. "Robots, Law and the Retribution-Gap." *Ethics and Information Technology* 18 (4): 299–309.

Duff, R. A. 1998. "Responsibility." In *Routledge Encyclopedia of Philosophy*. Taylor and Francis. https://www.rep.routledge.com/articles/thematic/responsibility/v-1.

Dworkin, G. 1981. "Voluntary Health Risks and Public Policy: Taking Risks, Assessing Responsibility." *Hastings Centre Report* 11: 26–31.

Goodin, R. E. 1987. "Apportioning Responsibilities." *Law and Philosophy* 6: 167–85.
Government Bill 1996/97:137 *Nollvisionen och det Trafiksäkra Samhället* [Vision Zero and the Traffic Safe Society]. Stockholm: Swedish Government.
Grill, Kalle, and Jessica Nihlén Fahlquist. 2012. "Responsibility, Paternalism and Alcohol Interlocks." *Public Health Ethics* 5 (2): 116–27.
Hansson, Sven Ove. 2007. "Hypothetical Retrospection." *Ethical Theory and Moral Practice* 10: 145–57.
Hart, H. L. A. 2008. *Punishment and Responsibility. Essays in the Philosophy of Law*. Second edition, with an introduction by John Gardner. Oxford: Oxford University Press.
Hevelke, Alexander, and Julian Nida-Rümelin. 2015. "Responsibility for Crashes of Autonomous Vehicles: An Ethical Analysis." *Science and Engineering Ethics* 21 (3): 619–30.
Honoré, Antony, and John Gardner. 2010. "Causation in the Law." In *Stanford Encyclopedia of Philosophy*, edited by Edward N. Zalta. https://plato.stanford.edu/archives/fall2019/entries/causation-law.
Hoover, K. D. 1990. "The Logic of Causal Inference." *Economics and Philosophy* 6: 207–34.
Hydén, Christer. 2019. "Speed in a High-Speed Society." *International Journal of Injury Control and Safety Promotion*. DOI: 10.1080/17457300.2019.1680566.
Kroes, Peter, and Peter-Paul Verbeek, eds. 2014. *The Moral Status of Technical Artefacts*. Springer: Dordrecht.
Lagnado, David A., and Tobias Gerstenberg. 2017. "Causation in Legal and Moral Reasoning." In *Oxford Handbook of Causal Reasoning*, edited by Michael R. Waldmann, 565–602. New York: Oxford University Press.
Machan, T. R. 1987. "Human Rights, Workers' Rights, and the 'Right' to Occupational Safety." In *Moral Rights in the Workplace*, edited by G. Ezorsky, 45–50. Albany, NY: State University of New York Press.
Mill, John Stuart. [1843] 1996. *A System of Logic*. In *Collected Works*, volume 7, edited by John M. Robson. Toronto: University of Toronto Press.
Mladenovic, Milos N., and Tristram McPherson. 2016. "Engineering Social Justice into Traffic Control for Self-Driving Vehicles?" *Science and Engineering Ethics* 22 (4): 1131–49.
Moore, Michael S. 1999. "Causation and Responsibility." *Social Philosophy and Policy* 16 (2): 1–51.
Moore, Michael S. 2009. *Causation and Responsibility: An Essay in Law, Morals, and Metaphysics*. New York, NY: Oxford University Press.
Noggle, Robert, and Daniel E. Palmer. 2005. "Radials, Rollovers and Responsibility: An Examination of the Ford-Firestone Case." *Journal of Business Ethics* 56 (2): 185–204.
Nyholm, Sven. 2018a. "Attributing Agency to Automated Systems: Reflections on Human–Robot Collaborations and Responsibility-Loci." *Science and Engineering Ethics* 24 (4): 1201–19.
Nyholm, Sven. 2018b. "The Ethics of Crashes with Self-Driving Cars: A Roadmap, II." *Philosophy Compass* 13 (7): e12506.

Rizzi, Dominick A., and Stig Andur Pedersen. 1992. "Causality in Medicine: Towards a Theory and Terminology." *Theoretical Medicine* 13: 233–54.

Rosencrantz, Holger, Karin Edvardsson, and Sven Ove Hansson. 2007. "Vision Zero—Is It Irrational?" *Transportation Research Part A: Policy and Practice* 41: 559–67.

Sparrow, Robert. 2007. "Killer Robots." *Journal of Applied Philosophy* 24 (1): 62–77.

Spurgin, E. W. 2006. "Occupational Safety and Paternalism: Machan Revisited." *Journal of Business Ethics* 63: 155–73.

van de Poel, I. 2011. "The Relation Between Forward-Looking and Backward-Looking Responsibility." In *Moral Responsibility*, edited by N. Vincent, I. van de Poel, and J. van den Hoven, 37–52. Springer: Dordrecht.

WHO. 2018. *Global Status Report on Road Safety 2018*. World Health Organization. https://www.who.int/violence_injury_prevention/road_safety_status/2018/en.

Wittgenstein, Ludwig. 1976. "Cause and Effect: Intuitive Awareness." *Philosophia* 6 (3): 409–25.

Chapter 5

The Ethics of Transitioning Toward a Driverless Future

Traffic Risks and the Choice Among Cars with Different Levels of Automation

Sven Nyholm

When I give lectures to students about the ethics of automated driving, I usually start by showing a video of a 2015 TED Talk given by Chris Urmson (Urmson 2015). At that time, Urmson was leading the self-driving cars project at Google, and his TED Talk is about as enthusiastic about the potential for what he calls "fully self-driving cars" as any presentation could be. Among the noteworthy things about that talk is that Urmson advocates in favor of a direct jump to fully self-driving cars, as opposed to first gradually working through increased levels of automation with different types of driver assistance systems. Another noteworthy thing is Urmson's confidence that within just a few years, fully self-driving cars will be so safe that they almost never get into accidents. Urmson ends his talk by showing a picture of his two sons. He says that within a little over five years, his oldest son will reach the age when he can get a driver's license. The goal of his self-driving cars project, Urmson concludes, is to make sure that his son won't need to learn how to drive. By then, fully self-driving cars should be available and be the only option worth considering.

When I give those lectures, I usually also show a short interview clip from around that same time, in which Elon Musk comments on the "Autopilot" feature in Tesla's Model S cars, which had been introduced around that time (Bloomberg 2014). That video is a very interesting contrast to Urmson's TED

Talk. Unlike the prototypes Google was developing around that time, Tesla's Model S with the autopilot feature was already available on the market, and Musk is much more careful in his claims about what automated driving can and cannot do. He specifically notes that Tesla calls what they offer an "autopilot" feature rather than "full autonomy" because Tesla wants to emphasize that there still has to be a person watching over what the car is doing, being ready at any time to take over the wheel if necessary. Notably, Musk does echo Urmson's prediction that full autonomation might be available and ready to go on the market within "five to six years." But he is more tentative in his statements than Urmson is in his. Musk says, among other things, that in order for law makers and the general public to accept fully automated cars, such cars would need to prove themselves to be much safer than regular cars. But it is worth noting that elsewhere—also around that same time—Musk predicted that when fully autonomous cars eventually do become much safer than regular cars, manual driving is likely to become illegal, given how risky it would then be in comparison to using fully autonomous cars (Hof 2015).

The video clips I have just described help to introduce the topic I will discuss in this chapter: namely, cars with different levels and types of automation, and the ethics of choosing among different kinds of cars. How should we assess choices people might make concerning what kinds of cars to use, given the assumption that cars with different levels and types of automation may be more or less safe, imposing different kinds of risks on others? For example, if some type of car—for example, a type of car with a high degree of automation—turns out to be much safer than other types of cars with other or no types of automation, would this create an ethical duty to switch over to using such cars? Or could there be any kinds of legitimate justifications for not making that switch? And should people have a right to make this choice? I will discuss some different perspectives on this cluster of questions and ultimately arrive at the following general suggestion: if some type of car with some form of automation turns out to be much safer than other types of cars, then this creates an ethical duty to either switch over to using this safer alternative or to use/accept additional safety precautions when using other forms of cars. For example, if a car with a potential for "full autonomy" would turn out to be safer than a car with some form of "autopilot" feature—to use the terminology from earlier—then it could reasonably be argued that there would be an ethical duty to either switch over to using the car that can operate in a fully autonomous mode or use added safety precautions when using the car with the autopilot feature.

Here is the plan of this chapter. I will start with some brief background about the idea of different "levels" of automation and comment on how this might be related to recent ideas within philosophical debates regarding how some technologies might or might not make humans "obsolete" with respect to the performance of certain tasks. I will then provide a little more background about discussions of traffic and driving within the philosophy of risk. Having done those things, I will first consider some interesting suggestions about the ethics of choosing between regular cars and autonomous cars made by Robert Sparrow and Mark Howard (2017) that to some extent echo the sentiments in the previously mentioned TED Talk by Chris Urmson. I will offer some criticisms of Sparrow and Howard's perspective and then consider an alternative suggestion from Jan Gogoll and Julian Müller (2020) about why, in their opinion, Elon Musk might be right that regular cars should become illegal once autonomous cars start outperforming them in terms of safety. I will end by discussing whether there could be an ethically acceptable way in which cars with less automation could still be used. In discussing that question, I will draw some inspiration from an interesting view about climate ethics presented by John Broome (2012) in another context. This will take me to the suggestion about the ethics of the choice among cars with different kinds of automation and safety levels briefly mentioned in the paragraph before this one.

LEVELS OF AUTOMATION AND PHILOSOPHICAL DEBATES ABOUT HUMAN OBSOLESCENCE

There are a few different ways in which it has been suggested that we can distinguish among different levels or types of automation in cars. The most well-known classification scheme is the one from the Society of Automotive Engineers (SAE). The SAE distinguishes among what they call six different "levels" of automation (SAE International 2021). It is worth noting that in the document in which the SAE suggests this classification scheme, they also recommend against using terminology such as "self-driving cars," "driverless cars," or "autonomous vehicles." Instead they suggest that it is more accurate and less misleading to talk about "driving automation systems" of different sorts (SAE International 2021). Their "levels of automation" associated with vehicles with or without any automation systems are the following:

- **Level 0: No Driving Automation:** The human driver performs all the tasks involved in driving conventional automobiles.
- **Level 1: Driver Assistance:** Some driving subtasks are taken over by the automation system in the vehicle.
- **Level 2: Partial Driving Automation:** Similar to level 1, but here more driving subtasks are taken over by the automation system.

- **Level 3: Conditional Driving Automation:** The automation system becomes able for certain time periods to perform all driving subtasks, but the human driver needs to be on standby and be ready to intervene.
- **Level 4: High Driving Automation:** Highly automated driving where there is a much less pronounced expectation that there will be any need or reason for the human driver to intervene.
- **Level 5: Full Driving Automation:** The car performs all the driving tasks, with no expectation that the human operator should take over any of the driving tasks.

To refer back to the introduction again, Tesla Model S cars with their "autopilot" feature would be either level 2 or 3 kinds of cars, whereas the Google cars that Urmson envisions in his TED Talk would be level 5 cars. Another note about these so-called levels of automation: given how abstractly they are defined, one can imagine different forms of technology being used to achieve them. And, moreover, one can also imagine cars on different levels of automation as potentially automating different driving subtasks. For example, some cars might feature more or less fully automated parallel parking capacities. Other cars might feature more or less fully automated highway driving. Other cars, in turn, might feature both those automation capacities but not the (full) automation of driving in urban areas, for example. Accordingly, automation in driving can take many different forms and be achieved by different kinds of technologies.

Discussions of these various forms of automation sometimes involve the claim that cars with these different types of automation will differ radically from each other in terms of their safety levels. It is sometimes claimed, for example, that expecting human drivers to perform only some, but not all, driving tasks might be highly risky—especially if this requires that human drivers be on standby, having to pay attention and potentially being asked by the system to suddenly take over (for example, Hevelke and Nida-Rümelin 2015). Human beings, it is then often said, have short attention spans; and they are also limited in their capacities to switch from paying attention to one thing (for example, a book they are reading) to another (for example, having to suddenly steer the car they are riding in) (Sparrow and Howard 2017). The goal, therefore, is sometimes said to be to take the human driver out of the driving equation. And some argue that this should be done as soon as possible because "humans are terrible drivers" (Urmson 2015).

This can be likened to a recent discussion within the ethics of human enhancement about whether some forms of technology might end up making humans or human interventions "obsolete." Robert Sparrow, for example, worries that if some human beings start using certain forms of human enhancement technologies, then other human beings, who do not use those technologies, might become obsolete (Sparrow 2019). Similarly, when philosophers and others discuss automated technologies and the future of work, this is another context in which worries are sometimes expressed that humans will become obsolete (Danaher 2019).

One interesting way of classifying different possibilities related to human obsolescence comes from work by John Danaher (Danaher in press). Danaher suggests that we should make two key distinctions. One is between whether humans will merely be perceived to be obsolete as opposed to actually becoming obsolete. Another distinction is between whether human beings are becoming altogether obsolete with respect to some broad set of tasks (for example, those associated with some profession) or whether human beings' skills are more narrowly obsolete with respect to some subset of a larger set of tasks (for example, some subset of the tasks associated with some profession).

That pair of distinctions can be interestingly related to the predictions people are making about future cars and automation in driving. Right now, some are certainly having the perception that human drivers will become obsolete: that their capacities and abilities will not be needed in the operation of a vehicle. This can be distinguished from the question of whether this will actually become the case and, if so, how soon that would happen. The other question is whether human input to driving will be broadly obsolete, so that there are no aspects of driving in different kinds of driving conditions where human input would be needed—or whether there would only be a more limited set of circumstances where the automation systems in the car would do a better job than a human driver.

SOME BACKGROUND ON THE PHILOSOPHY OF DRIVING RISKS

The risks related to driving and their ethical significance is not a topic that has been widely discussed in moral philosophy before automated driving became a hot topic. Typically, if driving was brought up as an example, it was said to be a kind of risky activity that, despite being risky, is widely considered to be morally acceptable. Sven Ove Hansson, for example, is a philosopher who has done a lot of important work in the philosophy of risk. He often brings up driving as an example of an ethically acceptable form of risk taking. According to Hansson, driving risks can be deemed morally acceptable for the reason that they are related to a human practice (driving) that is part of an equitable system of risk taking that works to the advantage of all or most members of society (Hansson 2003). Because it is good for us to have the opportunity to use cars, we can all, according to Hansson, accept that people are allowed to impose risks on each other as they do when they drive cars.

Of those who have discussed the ethics of driving specifically—that is, as opposed to bringing up driving merely as an example in more general discussions of risk—the authors of the very few articles that there are on the topic have tended to be highly critical of the risks we impose on each other when we drive cars. Douglas Husak, for example, argues in a 2004 article that driving is a much more morally problematic practice than both most regular people and moral philosophers tend to think (Husak 2004). Husak focuses in particular on what he calls "crash-incompatible vehicles"—viz. cars that impose more or less risk on other cars in case of crashes, for example, SUVs versus small cars. And Husak argues that the frivolous and unnecessary use of highly crash-incompatible cars ought to be considered unethical. In other words, if you drive around in a big SUV that would cause a lot of damage to other cars if there is a crash, and you have no good reason for driving around in this type of car other than that you (say) enjoy it, this is a morally questionable risk-imposing behavior, as Husak sees things.

Two other contributions to the ethics of driving risks that are relevant to the present discussion concern the types of precautions we could take—but that most people do not currently take—in order to make our driving of regular cars safer than it is. Kalle Grill and Jessica Nihlén-Fahlquist argue that because drinking alcohol before driving makes our driving much riskier, there is a strong moral case in favor of equipping all cars with alcohol interlocks (Grill and Nihlén-Fahlquist 2012). This way, people can only drive if and when they are sober, thereby reducing the risk of drunk driving leading to harmful or deadly accidents. In a similar vein, Jilles Smids discusses speed-regulating technologies of different kinds—"intelligent speed adaptation" (Smids 2018). He argues that there is a strong moral case in favor of equipping all cars with speed-limiting technologies that would make it either more difficult or impossible to disobey speed limits. Speeding, Smids argues, greatly increases driving risks, and it offends against democratically regulated rules. Moreover, cars can fairly easily and at reasonable cost be equipped with speed-limiting technologies. Therefore, according to Smids, there is a moral case in favor of mandatory speed-limiting technologies in all cars. Notably, many trucks are equipped with both alcohol locks and speed-limiting technologies, which makes the driving of long-haul truck drivers safer. According to Grill and Nihlén-Fahlquist, and to Smids as well, this should not just apply to trucks, but to all cars—and the reason is that society is under an ethical obligation to reduce the risks associated with driving.

I bring up these ideas about the ethics of driving risks from Hansson, Husak, Grill and Nihlén-Fahlquist, and Smids partly to give an idea of what was discussed in philosophical treatments of the ethics of driving risks before automated driving became a popular topic. But I also bring up these ideas because, as will become apparent later on, these various different ideas are highly relevant to the topic of this chapter, that is, the ethics of the choice among cars with different levels or types of automation.

SPARROW AND HOWARD ON THE ETHICS OF SELF-DRIVING CARS VERSUS REGULAR CARS

In a 2017 article called "When Humans Are Like Drunk Robots," Robert Sparrow and Mark Howard follow the lead of Chris Urmson in the previously mentioned TED Talk by positing that within the not-too-distant future, fully autonomous cars will become much safer than regular cars (Sparrow and Howard 2017). Sparrow and Howard also follow the lead of those who argue that partially automated cars—for example, level 2 and 3 cars—might involve significant risks, given that people might not be quick enough in their reactions to handle takeovers in risky driving scenarios. Therefore Sparrow and Howard accept the view I also associated with Urmson earlier according to which we should try to move directly from manually driven cars to fully automated cars, skipping any of the other steps in between with different levels or kinds of automation. Once fully autonomous cars are safe enough, Sparrow and Howard argue, human drivers will be comparable to "drunk robots": that is, human drivers will be poor substitutes for fully autonomous driving systems. In terms of the obsolescence that Sparrow himself worries about in other domains of the ethics of technology, it would be a good thing, Sparrow and Howard in effect argue, if human drivers would come to be actually and broadly obsolete with respect to all different possible tasks involved in operating vehicles.

Sparrow and Howard's discussion in their article also has something in common with the views expressed by Elon Musk that I brought up in the introduction. They argue that once self-driving cars become safer than regular cars, regular cars ought immediately to be prohibited. That is, we should not be permitted to drive regular cars anymore if and when self-driving cars become safer than regular cars. However, up until this happens—that is, up until self-driving cars are safer than regular cars—self-driving cars ought to be forbidden on public roads. As Sparrow and Howard see things, then, the choice between regular cars and automated cars—and fully automated cars in particular—should be determined by one thing and one thing only: namely, safety levels. They think that manually driven cars and fully automated cars are the only two sufficiently safe options. And out of the two, whatever is the safer option is the only option that ought to be afforded to people.

THREE WORRIES ABOUT SPARROW AND HOWARD'S VIEW

Sparrow and Howard's view has three main parts: first, the assumption that fully automated cars will fairly soon become much safer than regular cars in all kinds of traffic situations; second, the suggestion that we should not introduce any other kinds of partial automation before moving to full automation; and third, the suggestion that only regular cars should be used up until fully self-driving cars are safer than they are, at which point only fully self-driving cars should be permitted. This is an interesting overall perspective, but it is also a problematic one.

When it comes to the idea that fully self-driving cars are likely to become safer than regular cars across the board in all kinds of scenarios, and that this would happen fairly soon (for example, within a decade), this might be thought to be overly optimistic with respect to what the technology can do. David Mindell, an engineer and historian of technology at MIT, takes this perspective (Mindell 2015). Mindell argues that while full automation is technically possible in predicable and structured environments, full automation is much more difficult to achieve in many of the kinds of unpredictable and unstructured environments in which cars are driven. In general, automated robots tend to work best in controlled and predictable environments, where there is a limited set of tasks that the robots need to perform and where there are not too many people around to get in their way (Royakkers and van Est 2015). But the various different environments in which people like to drive their cars are not like this. Cars have to interact with other cars, with people inside and outside of the cars (for example, bicyclists and pedestrians), and with anything else that might appear on the road (for example, animals running across the road or branches falling from trees) (Nyholm and Smids 2020). Automated cars also need to be able to perceive and adequately classify things in their path and in the environment around them. This is a difficult challenge, especially in rough weather conditions. Moreover, driving on the highway is very different from driving, for example, in the middle of Amsterdam or any other complicated urban environment (Boffey 2019). Mindell discusses the example of driving in Boston in the winter: there might be lots of snow suddenly falling, and the environment is highly unstructured and unpredictable (Mindell 2015). According to Mindell, when we take these different kinds of considerations in mind, it should make us more skeptical of the idea that cars with full, level 5 automation will be safer than other kinds of cars with more or less automation across the board in all kinds of conditions and environments.

Another issue with the Sparrow and Howard view is that there are those who value driving and who will insist that they should have a right to drive cars. For example, let us briefly consider the "Human Driving Manifesto," written by Alex Roy, the founder of the "Human Driving Association." Here are some excerpts:

1: We are Pro-Human, in the pursuit of life, liberty, and freedom of movement . . .

2: We are Pro-Technology, but only as a means, not an end . . .

3: We are Pro-Safety, though a combination of improved driver education, deployment of Advanced Drivers Assistance Systems . . . and Parallel Automation.

4: We Support Raising Driver Licensing Standards . . .

[. . .]

6: We are Pro-Steering Wheel: No vehicle should be deployed without a steering wheel . . .

[. . .]

9: We Support Freedom of Movement and Traffic Neutrality . . .

[. . .]

12: We are Pro-Constitutional Amendment, creating a right to drive, within the limits of safety technologies that do not infringe upon our freedom of movement. (Roy 2018)

It is to be expected that more voices will join the "Human Driving Association" and insist upon the right and freedom to drive. What is interesting about this particular "manifesto" is that it shows an openness to added safety precautions, meaning that the manifesto is not a wholly unreasonable insistence that people be allowed to drive conventional cars whatever the cost to others might be. The question therefore arises whether the supposed ethical duty to switch to fully automated cars could be justifiably challenged by those who think they have a right to drive so long as they take added safety precautions that could help to make their driving as safe as possible (cf. Nyholm 2020, chapter four).

A third worry about the view suggested by Sparrow and Howard is that it is simply unrealistic and not feasible to suppose that we can switch directly from a car fleet with only regular cars to a car fleet with only fully automated cars. In September 1967, Sweden switched overnight from driving on the left-hand side to driving on the right of the road (Newsham 2016). That was possible. But the prospect of switching overnight—or even over a period as long as a year or a decade—from manually driven cars to only fully self-driving cars is not a realistic prospect. There are simply too many conventional cars everywhere. Any more realistic phasing out of manually driven conventional cars with little or no automation would have to happen over a longer period of time. Moreover, partial automation in cars is already being introduced. I already mentioned the case of the "Autopilot" feature of Model S Tesla cars. But there are also other car brands (such as BMW and Volvo) with different types of automation in the new cars that they are selling. So in terms of what is feasible and realistic, it is better to think of the introduction of automated cars in terms of the introduction of more or more different kinds of automation. This leaves us with a situation in which people who use cars face a choice among cars with different levels and types of automation. And it leaves ethicists discussing the ethics of that choice with a stronger reason to discuss it rather than only the choice between cars with no automation and cars with full automation.

THE RELEVANCE OF HANSSON'S IDEA OF AN EQUITABLE SYSTEM OF RISK TAKING THAT WORKS TO THE ADVANTAGE OF INDIVIDUALS EXPOSED TO RISK

When Jason Brennan discusses the ethics of vaccination in one of his articles, he makes use of the principle about acceptable risk imposition suggested by Sven Ove Hansson that I briefly introduced earlier (Brennan 2018). And Brennan makes an interesting comparison between driving and other risky practices we might engage in. In particular, Brennan assumes for the sake of his argument that there is some form of bomb that one could walk around with that would have the following property: the likelihood that the bomb would suddenly explode and end up killing somebody is equal or roughly equal to the likelihood that we might accidentally hit and kill somebody while driving. Brennan expects most of us to have an intuitive sense that while driving is permissible, risky though it might be, walking around with the bomb in this example is not acceptable, even though the risk level is the same. Why should this be so?

To explain this difference, Brennan refers to the principle of acceptable risk imposition that Hansson suggests (cf. Hansson 2003). In one case, the risky activity can be seen as being part of an equitable system of risk taking that works to the advantage of each person, whereas in the other case, the risky activity is not. Driving is part of such an equitable system of risk taking that works to the advantage of each person. But carrying around a bomb that might explode and kill somebody is not. That, Brennan suggests, is why these two activities that are stipulated to be equally risky differ in how morally acceptable they are.

In Brennan's own use of this line of reasoning, it is a step on the way to a conclusion about another subject matter: namely, the ethics of vaccination.[1] But in an interesting use of Brennan's line of reasoning that is more directly relevant for the main topic of this chapter, Jan Gogoll and Julian Müller apply it to the choice between manual driving and automated driving (Gogoll and Müller 2020). Using Brennan's types of reasoning, they argue that if and when automated driving becomes safer than manual driving, it removes the previous justifications we have had for viewing manual driving as being acceptable despite the risks it involves.

Gogoll and Müller argue as follows: when automated driving becomes safer than manual driving, this development weakens the argument for accepting regular driving that refers to how it is a part of a risk-imposing practice that works to the advantage of each person on whom a risk is imposed. It weakens that argument for the reason that there is suddenly an alternative practice—namely, the use of automated cars—that has the same advantages but is less risky for each person. Therefore the kind of reasoning Hansson introduces should here lead us to compare the different systems of risk imposition and ask which of the two is less risky while delivering the same or roughly equal benefits. And the suggestion is that the system whereby people use automated cars will soon become the winner.

This is certainly an interesting suggestion and a plausible argument. But what if people cannot afford, would be unwilling to, or would otherwise object to using automated cars rather than regular cars? Is there any argument we could make that would take into account the force of the argument presented by Gogoll and Müller, but that would somehow nevertheless leave open the option of using manually driven cars? Furthermore, as I suggested earlier, there is disagreement about how soon and whether it will be possible to create fully automated cars that function in automated modes in all different kinds of traffic situations and that would make human contributions to driving completely obsolete. In the more foreseeable future, it is likely that some sort of highly automated cars—but not 100 percent level 5 type of automation—is what is more realistic to think of as being what is available to car users. And if this is the case, there will presumably be different options available: cars with different kinds and levels of automation. Supposing that we agree with Gogoll and Müller that the introduction of more advanced forms of automation might change the moral landscape and put pressure on the justifiability of using certain kinds of cars, is there any way of interpreting any less automated forms of driving as still being acceptable or justifiable to some extent?

ARE THERE WAYS OF "OFFSETTING" RISKS INVOLVED IN DRIVING?

In a different context—namely, the ethics of how to respond to human-made climate change—the philosopher and economist John Broome presents a point of view that it is interesting to relate to the present topic of discussion (Broome 2012). Broome discusses the ethics of what we owe to future generations of people, and he is especially interested in whether our contemporary lifestyles are morally justifiable given the harm they might cause to future generations. According to Broome, current generations are wronging future generations by making the world less safe for them. Yet according to Broome, there are at least two morally justifiable ways in which we can respond to this problem of intergenerational harming. The most obvious thing to do is, of course, to curtail activities of ours that have an unacceptable carbon footprint. However, there is also another option that Broome claims is equally morally acceptable. We can also offset our carbon emissions. For example, we can plant trees. But there are also other things individuals can do to make sure that their own personal activities cannot be said to make any negative difference to how bad the environment will be for future generations. One example Broome discusses is buying stoves for people in developing countries whose current ways of cooking are very bad for the environment (Broome 2012, 14). The improvements that this could cause to the biosphere, Broome argues, would help to offset the negative impacts on the environment that we cause. Whatever the case might be when it comes to what the best ways of offsetting might be (planting trees, buying stoves for the relevant people, or whatever), the basic idea is that we can either reduce our own carbon footprint or neutralize it by making choices that help to offset it.

Say what you will about that argument in the context of climate ethics, I think an argument of a similar sort might be interesting to explore in the case of the ethics of the choice among cars with different types and levels of automation. My suggestion is as follows: suppose that there are extra precautions we could take that could help to "offset" any additional risks we might impose if we choose what are otherwise not the safest possible kinds of cars available to us. Suppose that this could help to make the choice of the less safe kind of cars more equal in safety levels to the safer alternatives available. Would this then not be a way of making it acceptable to drive the kinds of cars we might wish to drive?

On a more general level, the suggestion I have in mind is a principle that would say something along these lines: if a new technology is introduced into some domain and this new technology is safer than previously existing alternatives, then this creates an ethical duty to either (1) switch over to the safer alternative or (2) use or accept added safety precautions when using the older, less safe alternative (Nyholm 2020, chapter four). Applied to the case of cars with different types of automation, the result we would get is this: there is an ethical duty to either switch over to the types of more or less automated cars that are the safest, or to use/accept added safety precautions when using other, supposedly less safe kinds of cars, with other kinds of automation.

Note that in this formulation, I am not assuming that fully automated, level 5 cars are necessarily going to be safer than cars with some other form of automation (for example, level 3 or level 4 cars). I am instead making a general suggestion about whatever types or levels of automation might be the safest, which might eventually be level 5 cars or which might be some other form of cars. The idea is that when we choose among the different kinds of cars available to us, if there is a clear winner in terms of what imposes the least risk on others, then unless we take added precautions when using other kinds of cars, we might not have any valid moral justification for choosing those other forms of cars.

What kinds of added precautions could there possibly be? Well, we can here return to the ideas from Grill and Nihlén-Fahlquist about alcohol locks and the ideas from Smids about speed-limiting technologies that were briefly mentioned earlier (Grill and Nihlén-Fahlquist 2012; Smids 2018). Those would be examples of the kinds of added safety precautions that I have in mind. It could be argued that if one wants to continue driving a regular car after more automated forms of cars have been introduced, one might need to accept things such as alcohol locks or speed-limiting technologies in those regular cars—at least if more automated forms of cars have been proven to be safer than regular, manually driven cars. Such measures could help to at least partly offset the added risks that might be involved in a choice in favor of a less automated form of car if and when safer, more automated forms of cars become available.[2]

The principle of either switching over to a safer option or using added precautions if one uses an older, less safe option strikes me as a generally plausible principle. The plausibility of this principle can be illustrated with the following example. Imagine that some new form of medical treatment is introduced that is safer than some old form of treatment. But when you see your doctor, your doctor tells you that the old treatment will be used, despite the availability of the newer, safer treatment. It would appear that you have grounds for protesting here unless your doctor takes some added forms of precautions when administering the older, less safe form of treatment. But if extra precautions are used when the older, otherwise less safe form of treatment is administered, this would seemingly help to make up for, or compensate for, the unwillingness on the part of the doctor to use the newer, otherwise safer form of treatment. Similarly, if somebody was unwilling to switch over to some new, safer form of car that is safer because it is equipped with some specific form of automation, then they could potentially compensate for this if they use added precautions when using the form of car that they prefer driving.

Actually, this seems to be line with some of the points cited earlier from the "Human Driving Manifesto" from the Human Driving Association. Some of those points signal a willingness to use extra safety precautions, so long as individuals get to have their steering wheel and are given the option of driving their cars (Roy 2018). So this suggestion is more in line with the idea of giving people more options in the domain of driving than is a suggestion like Sparrow and Howard's.

How, it might also be asked, does the previously suggested principle relate to the Brennan and Hansson–inspired reasoning in Gogoll's and Müller's argument? It seems to me that the principle of either switching to some safer option or using added precautions is compatible with the principle of favoring systems of risk taking that work to the advantage of those on whom risks are imposed. Or rather, one does not have to choose between either accepting that type of principle or accepting the principle I suggest in this section. The two forms of reasoning about traffic risks could both be used in an overall view on how to reason about what choices people ought or ought not to be afforded within this domain.

CONCLUDING REMARKS

Opinions about if—and if so, how soon—fully automated driving will become safer than regular driving seem to differ greatly among the experts. At the point in time when this chapter is written, many experts seem to be a little more cautious in their assessments than they were a few years ago. For example, I mentioned Chris Urmson and the predictions he made about automated driving in his TED Talk in 2015 back when he was working for the Google self-driving cars project. In that TED Talk, as I mentioned earlier, Urmson indicated that he thought fully self-driving cars would be available on the market around the time that this book featuring this chapter appears. That did not happen. More recently, it has been reported that Urmson now has adopted a much more cautious view about when fully self-driving cars will be introduced. A 2019 piece in *The Economist* quoted Urmson as now being of the opinion that fully self-driving cars might slowly start to appear in a period of thirty to fifty years rather than the previously predicted timeframe of around five years.³

Whatever the case might be, it is reasonable to assume that car manufacturers will keep investigating and bringing to the market different kinds of automation in the new cars they are designing. And even before fully self-driving cars are introduced—and indeed even if they are never introduced—the ethical question of how to choose among cars with different types and levels of automation will still be one that those who use cars are confronted with. The principle that says to either go for the safest option or use added precautions when using less safe options will, I submit, be an ethically relevant principle during all of these different stages of development of different forms of automation in cars. It is an ethically relevant principle in this domain as well as in other domains where different types of technological solutions are on offer and the options differ in how safe or risky they are.

ACKNOWLEDGMENTS

Many thanks to the participants of my 2019 OZSW/4TU Ethics PhD course on the philosophy of risk, where I presented this material during one of the sessions. I benefited greatly from the feedback I received from the participants, which helped me very much in the preparation of this written version of the presentation I gave to them. Thanks also to Diane Michelfelder for her helpful feedback on the written version. My work on this chapter is part of the research program "Ethics of Socially Disruptive Technologies," which is funded through the Gravitation program of the Dutch Ministry of Education, Culture, and Science and the Netherlands Organization for Scientific Research (NWO grant number 024.004.031).

NOTES

1. As I am putting the finishing touches on this chapter, the COVID-19 pandemic is still ongoing. Notably, many of the arguments discussed in this chapter, not least Brennan's line of reasoning, are highly relevant to the ethics of how to safely transition out of a pandemic toward a postpandemic situation. However, because this volume is about the future of driving and not the ethics of pandemics, I set that topic aside here. But I discuss it elsewhere. See Nyholm and Maheshwari (in press).

2. Whether automotive manufacturers are going to want to produce cars with such features and whether those cars would be more or less expensive than other cars are issues I am making no assumptions about here. In this chapter, I am setting such practical issues aside and am asking what the right choice would be if such cars were

available at a reasonable cost and car users would thus face the choice of either using them or cars with significant forms of automation.

3. This was reported in an anonymous editorial called "Driverless Cars Are Stuck in a Jam," which is available on the website of *The Economist* here: https://www.economist.com/leaders/2019/10/10/driverless-cars-are-stuck-in-a-jam (accessed on November 10, 2019).

REFERENCES

Bloomberg. 2014. "Elon Musk on Tesla's Auto Pilot and Legal Liability." https://www.youtube.com/watch?v=60-b09XsyqU.

Boffey, Daniel. 2019. "Bikes Put Spanner in Works of Dutch Driverless Car Schemes." *The Guardian*. https://www.theguardian.com/world/2019/feb/13/bikes-put-spanner-in-works-of-dutch-driverless-car-schemes.

Brennan, Jason. 2018. "A Libertarian Case for Mandatory Vaccination." *Journal of Medical Ethics* 44 (1): 37–43

Broome, John. 2012. "The Public and Private Morality of Climate Change." *The Tanner Lectures on Human Values*. http://users.ox.ac.uk/~sfop0060/pdf/Tanner%20Lecture%20printed.pdf.

Danaher, John. 2019. *Automation and Utopia: Human Flourishing in a World Without Work*. Cambridge, MA: Harvard University Press.

Danaher, John. In press. "Technological Change and Human Obsolescence." *Techné: Research in Philosophy and Technology.*

Gogoll, Jan, and Julian Müller. 2020. "Should Manual Driving Be (Eventually) Outlawed?" *Science and Engineering Ethics* 26 (3): 1549–567.

Grill, Kalle, and Jessica Nihlén-Fahlquist. 2012. "Responsibility, Paternalism and Alcohol Interlocks." *Public Health Ethics* 5 (2): 116–27.

Hansson, Sven Ove. 2003. "Ethical Criteria of Risk Acceptance." *Erkenntnis* 59 (3): 291–309.

Hevelke, Alexander, and Julian Nida-Rümelin. 2015. "Responsibility for Crashes of Autonomous Vehicles: An Ethical Analysis." *Science and Engineering Ethics* 21(3): 619–30.

Hof, Robert. 2015. "Tesla's Elon Musk Thinks Cars You Can Actually Drive Will Be Outlawed Eventually." *Forbes*. https://www.forbes.com/sites/roberthof/2015/03/17/elon-musk-eventually-cars-you-can-actually-drive-may-be-outlawed.

Husak, Douglas. 2004. "Vehicles and Crashes: Why Is This Issue Overlooked?" *Social Theory and Practice* 30 (3): 351–70.

Mindell, David. 2015. *Our Robots, Ourselves: Robotics and the Myths of Autonomy*. New York: Viking.

Newsham, Gavin. 2016. "Urban Oddities: Cities Used to Do the Strangest Things . . ." *The Guardian*. https://www.theguardian.com/cities/2016/jul/19/urban-oddities-cities-strangest-things-baby-cages-airships-skyscrapers.

Nyholm, Sven, and Kritika Maheshwari. In press. "Offsetting Present Risks, Pre-empting Future Harms, and Transitioning towards a 'New Normal.'" In *Values for a Post-Pandemic Future*, edited by Matthew Dennis, Georgy Ishamaev, Steven Umbrello, and Jeroen van den Hoven. Berlin: Springer

Nyholm, Sven, and Jilles Smids. 2020. "Automated Cars Meet Human Drivers: Responsible Human-Robot Coordination and the Ethics of Mixed Traffic." *Ethics and Information Technology* 22 (4): 335–44.

Nyholm, Sven. 2020. *Humans and Robots: Ethics, Agency, and Anthropomorphism.* Lanham, MD: Rowman & Littlefield International.

Roy, Alex. 2018. "This Is the Human Driving Manifesto: Driving Is a Privilege, Not a Right. Let's Fight to Protect It." *The Drive*. https://www.thedrive.com/opinion/18952/this-is-the-human-driving-manifesto.

Royakkers, Lambèr, and Rinie van Est. 2015. *Just Ordinary Robots: Automation from Love to War*. Boca Raton: CRC Press

SAE International. 2021. "Taxonomy and Definitions for Terms Related to Driving Automation Systems for On-Road Motor Vehicles." https://www.sae.org/standards/content/j3016_202104.

Smids, Jilles. 2018. "The Moral Case for Intelligent Speed Adaptation." *Journal of Applied Philosophy* 35 (2): 205–21.

Sparrow, Robert. 2019. "Yesterday's Child: How Gene Editing for Enhancement will Produce Obsolescence—and Why It Matters." *American Journal of Bioethics* 19 (7): 6–15.

Sparrow, Robert, and Mark Howard. 2017. "When Human Beings Are like Drunk Robots: Driverless Vehicles, Ethics, and the Future of Transport." *Transport Research Part C: Emerging Technologies* 80: 206–15.

Urmson, Chris. 2015. "How a Self-Driving Car Sees the World." *Ted*. https://www.ted.com/talks/chris_urmson_how_a_driverless_car_sees_the_road/transcript.

Chapter 6

Stop Saying That Autonomous Cars Will Eliminate Driver Distraction

Robert Rosenberger

In early 2019, US senator Lamar Alexander, a Republican representing the state of Tennessee, offered a broad set of proposals for addressing climate change dubbed a "new Manhattan Project."[1] The main idea was to increase support for the search for innovative technological solutions, with an emphasis on nuclear energy and carbon capture. Additional proposals offered by Republicans at the time similarly were characterized by a deemphasis on regulation and an enthusiasm for potential technological fixes (for example, Wheeling,2019; Vorhees 2019). These proposals were a response to "the Green New Deal," an ambitious set of plans for addressing climate change developed by Democrats in the House of Representatives, which included a focus on drastically reducing greenhouse gas emissions, as well as a slate of clean air, food, employment, and justice-related initiatives (Friedman 2019). The contrasting visions of these proposals are showcased by their very different historical metaphors, one a reference to a concerted scientific effort to develop a particular new technology, the other to the expansive creation of government programs and projects. Independent of what we think of proposals for expansive action to address climate change like the Green New Deal, there is reason to remain suspicious of counterproposals offered by those with a history of climate denialism, as is the case for much of the GOP (Tang 2015). For evidence of this we need to look back only so far as the most recent Republican administration, under which then-president Trump attacked efforts to address climate change on multiple fronts, from drastically drawing down federal environmental regulations, to withdrawing from the Paris Climate Agreement, to systematically undermining climate science (Davenport and Landler 2019; see also Hansen 2021). My suggestion is that the Republican responses to the Green New Deal offer an instructive example from recent history of a particular kind of technology-focused argumentation.

There are two interesting aspects of the Republican proposals like the New Manhattan Project to note here. One, in contrast to the past positioning of much of the GOP, these proposals openly acknowledge that climate change *is actually happening*, that humans are at least in part to blame, and that something needs to be done. Two, they emphasize the promotion of general technological fixes that could emerge in the future over legislative restrictions that can be implemented now. As such, they can be interpreted as a particular kind of technological utopianism. They represent a specific conception of the nature of technological development: one that expects continuous positive advances, and one that expects those future advances to solve our contemporary problems. And they use this utopian technological outlook as a justification for dismissing specific action (in this case government regulations) on a contemporary problem. Don't do anything to solve the problem now because, perhaps with a push, technology will solve it in the future.

The question I would like to raise here is whether a similar kind of thinking will seep into the popular and political discourse on automated vehicles. The promise of self-driving cars is large. If they become fully (or even largely) adopted, then they could bring about massive changes to society, many of them potentially very positive changes. However, in the meantime there may grow a temptation to opt *not* to actively address particular current problems, and to instead wait for an expected autonomous vehicle revolution to solve them for us. Here, I urge us to resist this kind of utopian thinking around autonomous vehicles.

In what follows, in order to identify, articulate, and criticize this temptation, I first develop the notion of "spectatorial utopianism." I offer this as a label for the argument that we should refrain from addressing a contemporary concern based on a broad expectation of technological progress. Next, I consider a contemporary problem relating to the roadway: smartphones and distracted driving. We'll consider the science and theory on smartphone-induced driving impairment and consider the ways activists on this issue have urged for immediate action on this problem, including legislative restrictions and consciousness raising campaigns. Finally, we'll consider the potential for spectatorial utopian thinking on the topic of autonomous cars to derail efforts to address our contemporary problems relating to the road in general and to the problem of smartphone driving impairment in particular.

UTOPIAN SPECTATORSHIP

Richard Rorty often borrowed the term "spectatorial" from John Dewey to criticize those he thought were content to merely watch the world and theorize about it rather than work to change it for the better. Rorty in particular had his ire raised by the American academic left, which he argued had reclined into a position of mere critique rather than one engaged in the messy work of political organizing (1998). Here I'd like to borrow the term myself and bring this form of criticism to the philosophy of technology.

Let's use "spectatorship" as a term for calling out those positions that urge inaction on a pressing problem *based on one's abstract conception of the nature of technology*. That is, in the way I propose to use the term here, one's position can be critiqued as an example of mere spectatorship if it urges us to do nothing to change the status quo, and it justifies this inaction by appeal to a general philosophical account of technology.

For example, imagine someone who holds a totalizing dystopian view of technology. That is, imagine this person believes both that technology always (or at least usually) makes things worse and also that a future of more and more harmful technology is inevitable. For someone who holds this kind of determinist position, any new technology will be greeted as part of the problem. No version of technology will be seen as having the potential to make things better. And if they really think that some grim future is inevitable, then they may see any activism to be pointless. Of course dystopian views do not necessarily prompt inaction; I'm not here trying to implicitly cast shade on any particular broad philosophy. But were someone to hold a deterministic and dystopian position like the one in this paragraph, then it could be criticized as spectatorial.

Or, for example, we could imagine someone who sees technology to be inherently a nonplayer in human affairs. Again, while instrumentalist views do not necessarily lead to spectatorship, we can imagine instrumentalist positions that are open to this kind of critique. Imagine an instrumentalist who considers the artifacts of the world to be inherently beyond rebuke. In such a view, although a human user may be held responsible for things, this responsibility should never also extend to objects. We can imagine that this person might oppose any activism that urges for technological change (say, some kind of restriction on technology usage) on the basis of their "instrumentalist" view, that is, their abstract commitment to the idea that technology is inherently neutral. Bruno Latour's famous takedown of the National Rifle Association's (NRA) sloganeering is a good example (1999, 196). In opposition to any kind of gun control regulation in the United States, the NRA argues that guns don't kill people, people kill people. Latour exposes the shallow instrumentalism of this view, noting that it fails to recognize the change of agency that occurs as a person and a gun come together. "You are a different person with the gun in your hand" (Latour 1999, 179). The NRA's position can be criticized as spectatorial, content to merely bear witness to the status quo of tens of thousands of gun deaths annually in the United States, all justified (at least rhetorically, and perhaps disingenuously) on a general philosophical conception of the nature of technology.

Note that, in the way I am trying to use the term here, the accusation of spectatorship does not refer to just any position that calls for inaction or that defends the status quo. I want to reserve this term here—at least in its use in the philosophy of technology—only for those positions based largely on a general and abstract conception of technology that is then used to justify inaction on a particular concrete case. We could imagine many other calls for inaction or defenses of the status quo that would be of a different sort entirely. For example, if someone offers empirical information specific to the case, or a philosophical argument specific to the case, then this position would not be open to the allegation of advocating mere spectatorship. This is because the work of countering such a position should include addressing the actual case-specific information and/or arguments on offer rather than the dismissal of that position as too abstract and totalizing. That is, the allegation of spectatorship is reserved for those positions that seem to imply that, due to the very nature of technology, action against the technological status quo is always pointless or ill advised.

For example, in the introduction I pointed to one popular contemporary argument that I suggest embodies some of the worst impulses of technological spectatorship: the claim that we should not act on climate change based on the broad assumption that some technological fix will be invented in the future. But we can imagine arguments for doing nothing to fight climate change that are more concrete and specific. If someone were to offer a concrete prediction about the specific technology that will fix the environment, or offer specific information about how climate change is not a problem or is not actually occurring, then their position would not be one of mere spectatorship. Their urge for inaction on the environment, based on specific information and/or argumentation, would need to be countered with specific counter-information and counter-argumentation. We would need to expose their misinformation and specious argumentation.

Even still, we should be cautious: those defending spectatorial positions often attempt to dress them up as something more. Spectatorial arguments are often offered with a pretext of specificity. An argument might in spirit be largely one of spectatorship, and yet it may shallowly appeal to some kind of detail so as to not appear so transparently spectatorial. It is my opinion that this is the case for the appeals to the development of carbon capture and other specific technologies made by at least many of the conservative proposals for addressing climate change mentioned earlier. That argument is ultimately spectatorial: the main thrust of the position is that *we should not act* to restrict carbon emissions. The references to carbon capture technologies and the like are mainly present to provide an appearance of proactively addressing the problem.

This raises a second reason to remain cautious: spectatorial positions may be offered disingenuously. If someone has reasons to oppose concrete action to address a particular problem, and if they do not wish to reveal their real reasons for this opposition, then they may disingenuously claim that their motivation is based on a broad philosophical conception of the nature of technological progress. I believe this to be the case for our guiding example from recent history: the Republican counter-proposals to the Green New Deal mentioned at the start of the chapter. Without claiming to know the true motivations of any particular representative (for example, compared to other members of his party, Senator Alexander has long acknowledged manmade climate change), the vocal support these proposals received at the time from the party and from conservative media seems disingenuous; their real motivation was the opposition to concrete environmental regulations.

My goal here is to identify and articulate what I observe to be a particularly pervasive and pernicious form of technological spectatorship, and one to which thinking on autonomous vehicles may be especially susceptible: spectatorial utopianism.

In this form of spectatorship, a call for inaction is based on an assumption that technology will continue to advance in a way that will solve our problems. This species of utopianism recognizes the existence of the problems of the day, and it claims that technological fixes will emerge in the future. This kind of utopianism becomes "spectatorial" when it inspires one to *not* act to address those problems because of a commitment to a progress narrative with regard to technology.

The opening example of the recent Republican proposals to address climate change through a commitment to the future development of technological fixes is a paradigmatic example of spectatorial utopianism. These proposals are largely arguing against specific contemporary actions to reduce greenhouse emissions, and they are made on the basis of a broad conception of how problem-resolving technologies may advance.

Another example can be found in Ashley Shew's critiques of technoableist arguments found in the hype over exoskeleton development. According to Shew, rhetoric should be understood as "technoableist" when it ostensibly seeks to empower disabled people through the development of new technologies "while at the same time reinforcing ableist tropes about what bodyminds are good to have and who counts as worthy" (2020, 43). As one example, exoskeletons—that is, mechanical bodysuits that enhance mobility—are often hyped as a high-tech panacea for disabled people. But as Shew and other disability scholars and activists point out, exoskeleton advance remains a far-off future development, may forever be expensive to purchase and maintain, and in the end would still fail to even touch on many of the real problems that disabled people report to actually want to have addressed (for example, Ladau 2015; Sauder 2015; Earle 2019). And the most pernicious spectatorial utopian positions on this topic argue that we specifically should not focus on making spaces more accessible and should instead focus on developing exoskeleton technology.[2]

There are of course many ways to hold a utopian outlook without lapsing into spectatorship. One, utopianism must not automatically prompt inaction; one's commitment to a utopian perspective might instead inspire activism. That is, some utopians expect a better future due to technological advance and yet recognize that future will only be realized with active work toward it.

Two, someone may offer a utopian call against action on a current problem based on specific information or argumentation rather than an abstract commitment to general technological progress. For example, a person may be deeply involved in the development of a specific technology X that they believe will solve a pressing problem, and as someone with special knowledge they offer a specific prediction about when this technology can be expected to become available. This person may advise against, say, legislative action to address the problem on the argument that specific technology X will solve the problem on specific timetable Y. Were someone to disagree with this person, it will not be enough to simply dismiss their broad utopianism; the disagreement would need to be addressed at the level of these specifics. Will specific technology X actually solve the problem in the way advertised? If so, then is there a reason to be skeptical of timetable Y? Even still, it is important to remain cautious in these terms: spectatorial utopianism often attempts to pass itself off as if it is making a concrete argument about a specific technology and timetable and thus as something that can only be refuted in response to these specifics.

Because spectatorial utopian arguments have a veneer of engagement with the future, they can be particularly alluring and insidious. That is, because of their orientation toward the future, they can appear proactive, even as they are implicitly advocating inaction. I believe that it is possible that we will see this kind of argumentation arise based on claims about an upcoming future of autonomous vehicles. To consider just how dangerous this kind of argumentation can be, let's consider the case of smartphones and driver distraction.

SMARTPHONE DRIVING IMPAIRMENT

In the summer of 2018, the state of Georgia made a perhaps unexpected move: it outlawed handheld phone use for all drivers (Wickert 2018). The state's legislature was responding to a real and urgent problem. In recent years, the United States has seen a dramatic 14 percent spike in roadway fatalities, with Georgia the fifth highest in the nation, double the national rate of increase (Wickert 2017).[3] It is impossible to know for certain what caused the spike. But normal fluctuations (such as an increase of cars on the road) do not appear to be the culprit. The driver distraction of smartphones could be a factor. State legislators could no longer sit idly by.

The Georgia law, officially called "The Georgia Hands-Free Act," is good policy for several reasons. But it also has its limitations.

One reason laws like the Georgia Hands-Free Act are good policy is because it is easier to enforce than the state's former texting-only ban. Before the new law, Georgia did ban texting while driving, as almost all states in the United States also do. Research shows that texting while behind the wheel is enormously dangerous (for example, Drews et al. 2009; Yager 2012; Dingus et al. 2016; and for a meta-analysis of studies, see Caird et al. 2014). One oft-mentioned study by Rebecca L. Olsen and colleagues finds texting while driving to increase one's chances of an accident by twenty-three times (2009). I've seen this number lit up overhead on highway display signs in Georgia.

However, bans on texting have proven difficult to enforce. A police officer cannot tell if a driver is illegally texting or is instead touching a smartphone's screen in all the ways that remain legal, such as dialing or using apps. Hence one of the most important effects of a full handheld phone ban is that it deters texting while driving. Police officers can better enforce a full handheld ban, which encompasses texting, because they can simply pull over anyone holding a phone.

A second reason why a handheld smartphone ban is important is because it makes illegal the use of other handheld smartphone applications while driving. If texting while driving is dangerously distracting, then so is writing email, surfing the internet, and reading and writing social media posts. The State Farm insurance agency has conducted continuing surveys on smartphone usage and driving and has found that among US smartphone owners, more than a third (36 percent) admit to having used the internet while behind the wheel, a number which has continued to increase since their investigations began back in 2009 (State Farm 2021).

The third reason full handheld bans are important is both straightforward but also less intuitive: talking on the phone while driving is dangerous. It is understandable that texting while driving would be dangerous; texting involves a driver taking a hand off the wheel and, perhaps even more obviously important, taking their eyes off the road. But science has shown talking on the phone to be dangerously distracting, even though a driver keeps their eyes pointed forward. (You've probably also confirmed the science with your own observations of phone-using drivers on the road.) By studying drivers on test tracks and in driving simulators, by investigating police and hospital records, and by conducting surveys, scientists reveal the reality of smartphone-induced driving impairment.

Epidemiological studies of accident data and phone records have found crash risk to increase by up to four times when drivers talk on the phone (Redelmeier and Tibshirani 1997; McAvoy et al. 2005).[4] With test-track and simulator studies, scientists reveal correlations between phone conversation and behaviors that indicate reduced driving performance, such as poor speed control, increased reaction times, and failure to notice important cues (for a few recent simulator studies, see Farah et al. 2016; Haque et al. 2016; Choudhary and Velaga 2017; Calvi et al. 2018; Yan et al. 2018). Studies even find performance decrements to be comparable to drunk driving (for example, Strayer et al. 2006; Leung et al. 2012).

However, even though I strongly support legislation like the Hands-Free Georgia Act and see it as an important and life-saving step forward, we must simultaneously note a glaring drawback: it allows—and even encourages—hands-free phone usage. In the case of the Georgia law in particular, it's right there in the name. While handheld phone usage is now outlawed, hands-free usage is actively promoted. But this goes against the scientific evidence. In what is perhaps the central finding of more than two decades of research on mobile phones (and then smartphones) and driving, scientists find that both handheld and hands-free phone conversation result in the same dangerous decline in a driver's performance. The stack of research on this topic is copious enough to be the subject of a long series of meta-analyses and literature reviews (for example, McCartt et al. 2006; Lipovac et al. 2017; Caird et al. 2018).[5] As Jeff K. Caird and colleagues put it plainly in their recent meta-analysis of ninety-three experimental studies, "Because HH [handheld] and HF [hands-free] phone conversation produces similar driving performance costs, existing legislation that targets only HH phones may require reconsideration" (2018, 123).

More than only hands-free phone conversation, these technologies make possible voice interaction with smartphone applications, personal assistant programs, internet usage, and interface with vehicle systems. They also enable the audio reading and voice composition of text. Recent studies are beginning to show that this kind of hands-free interaction with the phone and the dashboard is very dangerous, comparable at times even to that of handheld texting, especially when the voice-translation technology is prone to error (for example, Strayer et al. 2013; Yager et al. 2013; He et al. 2014; Strayer et al. 2019; Zhang et al. 2019; Larsen et al. 2020; see Simmons et al. 2017 for a meta-analysis). As David L. Strayer and colleagues summarize, "The assumption that if the eyes were on the road and the hands were on the steering wheel then voice-based interactions would be safe appears to be unwarranted. Simply put, hands-free does not mean risk-free" (2013, 29).

DRIVER DISTRACTION AND THE FIELD OF AWARENESS

Issues of philosophy are introduced when we attempt to understand the nature of the distraction that stems from handheld and hands-free phone conversation. This is different from the distraction of looking away from the road to dial or to read or compose text (which is the most dangerous kind). When talking on the phone, drivers keep their eyes on the road. Something else is going on when a driver is distracted by the act of conversing on the phone.

One philosophical issue at the basis of the problem of smartphone driving impairment is the question of exactly how we should conceive of the human mind. Many hold the impression that if a person's eyes are open, then they must be seeing what's in front of them, at least under ordinary circumstances. And many also have the intuition that auditory perception—that is, listening to someone on the phone—shouldn't pose a distraction to the visual task of paying attention to the road. But as we've seen, the empirical research shows otherwise. How should we conceive of a mind that can be distracted in this manner?

I suggest that when we look at the language used in the empirical research papers, it is possible to abstract a default conception of the distracted mind at work in these studies. In my own work, I have developed an alternative way to understand how smartphone distraction operates.

Empirical research on distracted driving is mainly performed in the field of cognitive science. When these scientists report their findings, they often couch them in terms of a driver's inherently limited quantity of cognitive resources. The postulation is that a driver's mind cannot safely handle the two cognitively demanding tasks of driving and using the phone at the same time because a mind is inherently limited in its total stock of "cognitive resources," or "attention," or "information processing capacity" (see esp. Rosenberger 2012). A helpful term has come out of this discussion: cognitive distraction. In contrast to the visual distraction of looking away from the road (say, at a text message), and the manual distraction of taking a hand off the wheel (say, to dial), there is also a form of cognitive distraction that occurs as a driver's mind is taken off the road. This term has proven to be useful, finding its way into the rhetoric of many anti-driver-distraction consciousness raising campaigns, including those of the World Health Organization and the National Safety Council (WHO 2011; NSC 2012).

Across a series of papers, I have used ideas from the philosophy of technology to develop an alternative explanation of smartphone driving impairment (for example, Rosenberger 2012; 2013; 2015; 2019). I strongly agree that the data show smartphone usage while driving to be a dangerous activity, and I agree too that it should be better regulated. But I approach the interpretation of these data through a different conception of the human mind.

In particular, I've developed ideas from the "postphenomenological" perspective to give an account of the experience of phone usage.[6] Rather than take up a mechanistic conception of the human mind as something that processes particular quantities of resources, I try to describe what it's like to be on the phone. The two central ideas I use are a conception of "bodily-perceptual habituation" and a user's "field of awareness."

My suggestion is that while someone is on the phone, the experience is at least sometimes quite immersive. The practical totality of what the phone user is experiencing in that immersive moment (that is, the totality of their field of awareness) is composed by the content of the phone conversation and the presence of the person on the other end of the line. And because phone usage is an everyday activity, this specific immersive organization of awareness—this organization in which the near entirety of what is experienced is composed by the phone conversation—becomes sedimented within bodily-perceptual habituation. That is, because phone usage is so normal, so much a part of everyday life, so steeped in routine behavior, one may be at times pulled by habit into a kind of experiential immersion, and this can happen almost automatically. In the context of typical phone usage, this is fine. It is perhaps often even preferable. But for the special circumstances of using the phone while driving, this can be dangerous. Under this account, smartphone driving impairment occurs as a driver's overall field of awareness is pulled away from the road and into a composition of experience centered mainly on the conversation.[7] Even if you commit yourself to maintaining a strong primary focus on the road ahead as you drive and talk on the phone, your nondriving phone-related habits may creep in and subtly pull your focus from the road, perhaps as the conversation becomes riveting, or perhaps as the driving gets boring.

In any case, whether we adopt a theory of inherent cognitive limitations, or one of learned bad habits of immersive perception, or even some combination of the two, it is important to refine our explanations of smartphone-induced driving impairment. The development of theory on this topic is important because we need to devise whatever rhetorical strategies we can to help convince drivers to refrain from using the phone while driving. Although, as reviewed earlier, the empirical researchers have shown phone usage to dangerously degrade driving performance, many drivers are still not getting the message.[8] The law struggles to keep up with the latest technological advancements. And the big automotive and telecommunications companies are racing to cram as many "connectivity" features into our dashboards and mobile devices as possible.

This topic can thus be seen as an urgent combination of cutting-edge scientific research, engaged philosophical argumentation, political battles over the law, consciousness-raising activism, and the continued high-speed advance of two of today's most influential technologies, the car and the smartphone. And lives are on the line.

UTOPIANISM AND AUTOMATED CARS

The problem of smartphone-induced driving impairment is an interesting one to consider in light of the arrival of autonomous vehicles. This is because were our roadways to become fully automated, the problem would be dissolved. That is, if all cars drive themselves without the need for human driver input or supervision, then driver distraction would not be an issue at all.

I look forward to such a future. I am convinced that if the roadways become fully automated, then they would be much safer. Human driver error and irresponsibility appear to me to be to blame for much of the risk of the road. And that form of risk would be eliminated by the advent of cars that entirely drive themselves. Sure, computerized driving systems will not prove perfectly safe. But I strongly intuit that, were our roadways to become completely automated, the net roadway safety will greatly increase. And I suspect that many share my intuition.

It is this exact intuition that makes it possible that a dangerous spectatorial utopian thinking could enter our discourse on roadway safety and planning generally, and smartphone driver distraction in particular. The spectatorial utopian temptation may be to argue that because an approaching future of automated vehicles will solve the problem of driver distraction, we are thus released from any imperative to address the problem now.[9] Such an argument is utopian in that it relies on a general expectation of future technological advance that will solve our problems for us. It is spectatorial in that serves as a justification for inaction, and we can even imagine a version of this argument offered against a specific call for action, such as a call for greater regulation. And it is specious.

It is specious because it relies on the poorly supported assumption that a world of fully automated roadways is on the very near horizon. This is not a safe assumption for a number of reasons.

One, although the hype may imply otherwise, it is not certain that automated cars are actually upon us. It is true that a few expensive highly automated ones are available today, and some models are currently under scientific study. But it is not a certainty that automated vehicles will be widely available to the mass consumer in the immediate future. They might be! But it is a mistake to fail to address a pressing safety problem that *is* occurring now on the promise of a solution that only *might* be immediately forthcoming.

Two, even if automated cars are approaching on the near horizon, they will not necessarily soon be fully autonomous. That is, it is likely that the first phase of automated vehicles available and affordable to the everyday driver will only be partially automated. For example, the Society for Automotive Engineers has developed a taxonomy of the kinds of automation that we can expect for cars, one that has been taken up in US government documents (SAE 2018). A crucial jump occurs between what they identify as levels 2 and 3, with the former including a human driver that performs some aspects of the driving task, and the latter calling for the human driver to participate only in moments when intervention is requested by the automated system. This specific task of level 3 autonomous car usage—in which a person is asked to continuously diligently supervise an automated vehicle and stay ready to intervene in the driving at any moment, and yet otherwise not perform any of the actual driving—is beset with its own problems. As we've seen in the research and theory on smartphones and distracted driving, it is difficult for some drivers to stay focused even when they are in the middle of the task of actually driving the car. It may prove too much to ask drivers to maintain continual focus when the car is otherwise entirely driving itself. The growing stack of empirical research has so far yielded widely varying findings in driver "takeover times," that is, the time it takes for a driver to take control of the vehicle after a request by the system has been made.[10] In any case, we should not shirk our responsibility to address the problem of driver distraction on the expectation that cars will soon drive themselves if we cannot be sure that they will soon be fully autonomous.

Three, even if the technology for full automation were to become soon available, that would not ensure immediate full adoption. That is, even if we soon have the technological ability to make fully autonomous vehicles widely available, we still cannot assume that all drivers, or all kinds of drivers, will be immediately replaced by automation. We can imagine that particular sectors of drivers—say, truck drivers, or cab drivers—may become supplanted more quickly and more systematically than others. We can imagine too the possibility of a social resistance emerging in which some people decide not to cede the driving duty over to automation. Some time may need to pass before incentives for yielding driving control to automation (such as insurance incentives, or a market in which nonautomated cars become expensive) outweigh the individual choices of resisters. So even were we to stipulate that full automation is here now, if we cannot guarantee something near full adoption, then we should not abandon contemporary efforts to curb distracted driving.

Four, the project of fully automating our roadways may not reduce to the two factors of driver choices and the technological development of automated cars. It may turn out to also involve larger social and political factors, such as advancements in law or public infrastructure. It is possible that the march toward fully automated roadways will be slowed by legal questions around, say, who should be held responsible for damages caused by self-driving cars. And it may turn out to be the case that optimized automated roadway systems may call for different roads themselves rather than only differences made to vehicles that drive on otherwise traditional roads. If so, then it is easy to imagine that not all roadways will be updated at the same rate. The point here is that even if the technology for fully automated cars is soon widely available and even if there is the social will for full adoption, the shift to autonomous roadways could still be slowed by factors beyond the cars and drivers themselves.

With these various reasons in mind, we should regard with suspicion any claims that we should set aside efforts to address distracted driving based on an expectation that the issue will soon be made moot by automated cars. In addition to the potential speciousness of any version of this kind of argument we may encounter, we should also remain on the lookout for their potential disingenuousness. That is, some of those offering spectatorial utopian arguments against the regulation of smartphones while driving may be masking their deeper motivation: their own economic benefit of smartphone-using drivers. The increasing restriction of the use of smartphones while driving comes at the financial disadvantage of many powerful actors, including businesses who expect their workers' daily commute to become an extension of time in the office, the automotive industry that has been aggressively peddling dashboard infotainment systems, and the telecommunications industry that stands to lose an entire class of customer: those behind the wheel.

CONCLUSION

The arrival of automated cars would bring large-scale changes to society. And the hype around them would have us believe that such a future is fast barreling down upon us. With those two things in mind, we can expect forms of spectatorial utopian thinking to show up regarding any number of issues. Everything from issues traffic congestion, to mass transit infrastructure improvements, to the regulation of gig economy ridesharing services could be argued as unimportant to address *now* because a future of automated cars that will change everything is coming *soon*. Whether we are actually racing or inching toward that future, and whether that future is inevitable or only seemingly so, we should not use it as an excuse to avoid confronting our current problems.

The issue of smartphone driving impairment is only perhaps the most extreme one susceptible to spectatorial utopian thinking on autonomous vehicles. It is extreme because it is at once a controversial and urgent problem and at the same time one that would be completely obviated by the advent of the full adoption of entirely automated roadways. We should not succumb to the temptation of putting off addressing this important issue in the hope that new technology will simply make it go away. It is true that the arrival of fully automated cars would solve the problem of driver distraction. But we should stop saying it.

NOTES

1. See https://www.alexander.senate.gov/public/index.cfm/2019/3/one-republican-s-response-to-climate-change-a-new-manhattan-project-for-clean-energy-10-grand-challenges-for-the-next-five-years.

2. For example, Zoltan Istvan has recently made this exact argument. And it is consistent with at least some of the thinking coming out of transhumanist circles that sees disability as something to be surpassed rather than accommodated (see Istvan 2015; Eveleth 2015).

3. The recent national spike may have seen its peak at a 14 percent increase in 2014 through 2016, then dropping only 1 percent in 2017 (National Center for Statistics and Analysis 2017; 2018). It can be noted that according these US Department of Transportation reports, only a tiny percentage of these deaths can be attributed to smartphone usage. However, as groups like the National Safety Council have long insisted, smartphone-related driving incidents are vastly undercounted. The NSC in particular goes so far as to claim that around 25 percent of all traffic accidents involve smartphone usage (National Safety Council 2015).

4. A more conservative meta-analysis of epidemiological data by Rune Elvik found phone users to be 2.86 times more likely to be involved in a crash (2011).

5. For review and theory on the difference between phone conversation and passenger conversation, see Rosenberger 2019. For some important contradictory findings to the claim that hands-free phone use is dangerous, see the research associated with naturalistic in-car camera studies (esp. Dingus et al. 2016).

6. For introductions to postphenomenology, see Ihde 2009; Rosenberger and Verbeek 2015; Aagaard 2017.

7. Not all of my colleagues in postphenomenology and the philosophy of technology are convinced by my account here. For our debate, see the special issue 18(1–2) of the journal *Techné: Research in Philosophy and Technology* published in 2014.

8. This is not entirely surprising because studies also reveal that drivers are not especially good at gauging their own level of smartphone-based distraction (for example, Horrey et al. 2008; Sanbonmatsu et al. 2016; Terry and Terry 2016).

9. Technology critics sometimes flirt with spectatorial utopian argumentation against the regulation of smartphones while driving on the basis of an expected future of autonomous vehicles (for example, Manjoo 2010; Oremus 2016). And it has become a mainstay of telecommunications industry lobbying efforts to resist distracted driving regulation (for example, Levin 2017). For example, the Consumer Technology Association (a major trade organization representing everyone from AT&T, to Apple, to Google) released a statement criticizing Obama administration guidelines for the design of mobile devices and their pairing with dashboard systems. They argued that instead "we encourage NHTSA to rethink its approach on this issue, work with innovators to bring technology solutions to driver and focus on areas within its jurisdiction—bringing self-driving vehicles to market and eliminating the majority of roadway deaths." See www.businesswire.com/news/home/20161123005566/en/NHTSA-Guidelines-Jeopardize-Continued-Market-Driven-Innovations-to-Enhance-Driver-Safety-Says-CTA.

10. For a meta-analysis of 128 studies, see Zhang et al. 2019.

REFERENCES

Aagaard, J. 2017. "Introducing Postphenomenological Research: A Brief and Selective Sketch of Postphenomenological Research Methods." *International Journal of Qualitative Studies in Education* 30 (6): 519–33.

Caird, J. K., K. A. Johnston, C. R. Willness, M. Asbridge, and P. Steel. 2014. "A Meta-Analysis of the Effects of Texting on Driving." *Accident Analysis & Prevention* 71: 311–18.

Caird, J. K., S. M. Simmons, K. Wiley, and K. A. Johnston. 2018. "Does Talking on a Cell Phone, with a Passenger, or Dialing Affect Driving Performance? An Updated Systematic Review and Meta-Analysis of Results." *Human Factors* 60 (1): 101–33.

Calvi, A., A. Benedetto, and F. D'Amico. 2018. "Investigating Driver Reaction Time and Speed During Mobile Phone Conversations with a Lead Vehicle in Front: A Driving Simulator Comprehensive Study." *Journal of Transportation Safety & Security* 10 (1–2): 5–24.

Choudhary, P., and N. R. Velaga. 2017. "Modelling Driver Distraction Effects Due to Mobile Phone Use in Reaction Time." *Transportation Research Part C* 77: 351–65.

Davenport, C., and M. Landler. 2019. "Trump Administration Hardens Its Attack on Climate Science." *New York Times*, May 27. https://www.nytimes.com/2019/05/27/us/politics/trump-climate-science.html.

Dingus, T. A., F. Guo, S. Lee, J. F. Antin, M. Perez, M. Buchanan-King, and J. Hankey. 2016. "Driver Crash Risk Factors and Prevalence Evaluation Using Naturalistic Driving Data." *PNAS* 113 (10): 2636–41.

Drews, F. A., H. Yazdani, C. N. Godfrey, J. M. Cooper, and D. L. Strayer. 2009. "Text Messaging During Simulated Driving." *Human Factors* 51 (5): 762–70.

Earle, J. 2019. "Cyborg Maintenance: Design, Breakdown, and Inclusion." In *Design, User Experience, and Usability*, edited by A. Marcus and W. Wang, 47–55. Springer.

Elvik, R. 2011. "Effects of Mobile Phone Use on Accident Risk: Problems of Meta-Analysis When Studies Are Few and Bad." *Transportation Research Record* (2236): 20–26.

Eveleth, R. 2015. "The Exoskeleton's Hidden Burden." *The Atlantic*, August 7. https://www.theatlantic.com/technology/archive/2015/08/exoskeletons-disability-assistive-technology/400667/.

Farah, H., S. Zatmeh, T. Toledo, and P. Wagner. 2016. "Impact of Distracting Activities and Drivers' Cognitive Failures of Driving Performance." *Advances in Transportation Studies* 2016 Special Issue (1): 77–82.

Friedman, L. 2019. "What Is the Green New Deal? A Climate Proposal Explained." *New York Times*, February 21. https://www.nytimes.com/2019/02/21/climate/green-new-deal-questions-answers.html.

Gliklich, E., R. Guo, and R. W. Bergmark. 2016. "Texting While Driving: A Study of 1211 U.S. Adults with the Distracted Driving Survey." *Preventative Medicine Reports* 4: 486–89.

Hansen, S. 2021. "Republicans Claim Rejoining Paris Climate Accords Will Cost American Jobs, But Here's Really What's Happening." *Forbes*, January 21. https://www.forbes.com/sites/sarahhansen/2021/01/21/republicans-claim-rejoining-paris-climate-accords-will-cost-american-jobs-but-heres-whats-really-happening/?sh=77f184fbf4f5.

Haque, M. M., O. Oviedo-Trespalacious, A. K. Debnath, and S. Washington. 2016. "Gap Acceptance Behavior of Mobile Phone-Distracted Drivers at Roundabouts." *Transportation Research Record* (2602): 43–51.

He, J., A. Chapparo, B. Nguyen, R. J. Burge, J. Crandall, B. Chapparo, R. Ni, and S. Cao. 2014. "Texting While Driving: Is Speech-Based Text Entry Less Risky Than Handheld Text Entry?" *Accident Analysis and Prevention* 72: 287–95.

Horrey, W. J., M. F. Lesch, and A. Garabet. 2008. "Assessing the Awareness of Performance Decrements in Distracted Drivers." *Accident Analysis and Prevention* 40: 675–82.

Ihde, D. 2009. *Postphenomenology and Technoscience: The Peking University Lectures*. Albany, NY: SUNY.

Istvan, Z. 2015. "In the Transhumanist Age, We Should Be Repairing Disabilities, Not Sidewalks." *Motherboard*, April 3. https://www.vice.com/en/article/4x3pdm/in-the-transhumanist-age-we-should-be-repairing-disabilities-not-sidewalks.

Ladau, E. 2015. "Fix Discriminatory Attitudes and Broken Sidewalks, Not Humans." *Motherboard*, April 8. https://motherboard.vice.com/en_us/article/d73947/fix-discriminatory-attitudes-and-broken-sidewalks-not-humans.

Larsen, H. H., A. N. Scheel, T. Bogers, and B. Larsen. 2020. "Hands-Free But Not Eyes-Free: A Usability Evaluation of Siri While Driving." *2020 Conference on*

Human Information Interaction and Retreival (CHIIR '20). New York: ACM. doi.org/10.1145/3343413.3377962.

Latour, B. 1999. *Pandora's Hope: Essays on the Reality of Science Studies.* Cambridge, MA: Harvard University Press.

Leung, S., R. J. Croft, M. L. Jackson, M. E. Howard, and R. J. McKenzie. 2012. "A Comparison of the Effect of Mobile Phone Use and Alcohol Consumption on Driving Simulation Performance." *Traffic Injury Prevention* 13 (6): 566–74.

Levin, M. 2017. "Don't Drive Distracted, Wireless Industry Says, But Safety Advocates Want More Than Talk." *Fair Warning*, August 30. https://www.fairwarning.org/2017/08/wireless-companies-duck-responsibilities-distracted-driving/.

Lipovac, K., M. Đeric', M. Tešic', Z. Andric', and B. Maric'. 2017. "Mobile Phone Use While Driving—Literary Review." *Transportation Research Part F* 47: 132–42.

McAvoy, S. P., M. R. Stevenson, A. T. McCartt, M. Woodward, C. Haworth, P. Palamara, and R. Cercarelli. 2005. "Role of Mobile Phones in Motor Vehicle Crashes Resulting in Hospital Attendance: A Case-Crossover Study." *BMJ* 331: 428–33.

McCartt, Anne T., Laurie A. Hellinga, and Keli A. Bratiman. 2006. "Cell Phones and Driving: Review of Research." *Traffic Injury Prevention* 7 (2): 89–106.

National Center for Statistics and Analysis. 2017. *2016 Fatal Motor Vehicle Crashes: Overview.* Traffic Safety Facts Research Note. Report No. DHT HS 812 456. Washington, DC: National Highway Traffic Safety Administration.

National Center for Statistics and Analysis. 2018. *2017 Fatal Motor Vehicle Crashes: Overview.* Traffic Safety Facts Research Note. Report No. DOT HS 812 603. Washington, DC: National Highway Traffic Safety Administration.

National Safety Council. 2012. "Understanding the Distracted Brain: Why Driving While Using Hands-Free Cell Phones Is Risky Behavior." White Paper, April. https://www.nsc.org/road-safety/safety-topics/distracted-driving/distracted-brain.

National Safety Council. 2015. "Cell Phones Are Involved in an Estimated 27 Percent of All Crashes, Says National Safety Council." Press Release, NSC.org, May 18. http://www.nsc.org/Connect/NSCNewsReleases/Lists/Posts/Post.aspx?ID=9.

Manjoo, F. 2010. "Don't Worry, the Robot's Driving: Texting on the Road Is Dangerous. The Solution: Self-Driving Cars." *Slate*, October 12. https://slate.com/technology/2010/10/texting-on-the-road-is-dangerous-the-solution-self-driving-cars.html.

Olsen, R. L., R. J. Hanowski, J. S. Hickman, and J. Bocanegra. 2009. "Driver Distraction in Commercial Vehicle Operations." Report No. FMCSA-RRR-09-042. US Department of Transportation. Federal Motor Carrier Safety Administration. Form DOT F 1700.7 (8-72).

Oremus, W. 2016. "Self-Driving Cars Are Getting Better. Are Human Drivers Getting Worse?" *Slate*, October 6. https://slate.com/technology/2016/10/human-car-fatalities-are-rising-are-self-driving-cars-the-solution.html.

Redelmeier, D. A., and R. J. Tibshirani. 1997. "Association Between Cellular Telephone Calls and Motor Vehicle Collisions." *New England Journal of Medicine* 336: 453–58.

Rorty, R. 1998. *Achieving Our Country: Leftist Thought in Twentieth-Century America.* Cambridge, MA: Harvard University Press.

Rosenberger, R. 2012. "Embodied Technology and the Problem of Using the Phone While Driving." *Phenomenology & the Cognitive Sciences* 11 (1): 79–94.

Rosenberger, R. 2013. "The Problem with Hands-Free Dashboard Cell Phones." *Communications of the ACM* 56 (4): 38–40.

Rosenberger, R. 2015. "Driver Distraction from Mobile and Wearable Computer Interface." *IEEE Technology & Society Magazine* 34 (4): 88–99.

Rosenberger, R. 2019. "The Experiential Niche: Or, On the Difference Between Smartphone and Passenger Driver Distraction." *Philosophy & Technology* 32 (2): 303–20.

Rosenberger, R., and P.-P. Verbeek. 2015. "A Field Guide for Postphenomenology." In *Postphenomenological Investigations: Essays on Human-Technology Relations*, edited by R. Rosenberger and P-P. Verbeek, 9–41. Lanham, MD: Lexington Books/Rowman & Littlefield Press.

Sanbonmatsu, D. M., D. L. Strayer, F. Biondi, A. A. Behrends, and S. M. Moore. 2016. "Cell-Phone Use Diminishes Self-Awareness of Impaired Driving." *Psychonomic Bulletin & Review* 23: 617–23.

Sauder, K. 2015. "When Celebrating Accessible Technology Is Just Reinforcing Ableism." crippledscholar.com, July 4. https://crippledscholar.com/2015/07/04/when-celebrating-accessible-technology-is-just-reinforcing-ableism/.

SAE International. 2018. *Taxonomy and Definitions for Terms Related to Driving Automation Systems for On-Road Motor Vehicles* (J3016_201806). Warrendale, PA: SAE International.

Shew, A. 2020. "Ableism, Technoableism, and Future AI." *IEEE Technology & Society Magazine* 31 (1): 40–50, 85.

Simmons, S. M., J. K. Caird, and P. Steel. 2017. "A Meta-Analysis of In-Vehicle and Nomadic Voice-Recognition System Interaction and Driving Performance." *Accident Analysis and Prevention* 106: 31–43.

State Farm. 2021. "Distracted While Driving." April. https://newsroom.statefarm.com/why-did-89-of-drivers-choose-to-engage-in-distracted-driving-behaviors-in-2020/.

Strayer, D. L., F. A. Drews, and D. A. Crouch. 2006. "A Comparison of the Cell Phone Driver and the Drunk Driver." *Human Factors* 48 (2): 381–91.

Strayer, D. L., J. M. Cooper, J. Turrill, J. Coleman, N. Medeiros-Ward, and F. Biondi. 2013. "Measuring Cognitive Distraction in the Automobile." Washington, DC: AAA Foundation for Traffic Safety. www.aaafoundation.org/sites/default/files/MeasuringCognitiveDistractions.pdf.

Strayer, D. L., J. M. Cooper, R. M. Goethe, M. M. McCarty, D. J. Getty, and F. Biondi. 2019. "Assessing the Visual and Cognitive Demands of In-Vehicle Information Systems." *Cognitive Research: Principles and Implications* 4: 18. Doi.org/10.1186/s31235-019-0166-3.

Tang, V. 2015. "Here Are All The Senators Who Do and Don't Believe in Human-Caused Climate Change." Wired.com, January 21. https://www.wired.com/2015/01/senators-dont-believe-human-caused-climate-change/.

Terry, C. P., and D. L. Terry. 2016. "Distracted Driving Among College Students: Perceived Risk Verses Reality." *Current Psychology* 35: 115–20.

Vorhees, J. 2019. "The Next Partisan Climate Battle Will Be About *When* to Act." *Slate*, March 29. https://slate.com/news-and-politics/2019/03/mike-lee-tauntaun-climate-change-alexandria-ocasio-cortez.html.

Wheeling, K. 2019. "Will Republican Climate Change Proposals Work?" *Pacific Standard*, May 10. https://psmag.com/environment/will-republican-climate-change-proposals-work999999.

Wickert, D. (2017). "Georgia Motor Vehicle Deaths Jump By A Third in Two Years." *Atlanta Journal-Constitution*, 2/15/2017.https://www.ajc.com/news/local/georgia-motor-vehicle-deaths-jump-third-two-years/JUpDheU8eFb3GJlrobSn4I/

Wickert, D. (2018). "Safety Advocates: Georgia's Distracted Driving Law A 'First Step.'" *Atlanta Journal-Constitution*, 6/29/2018. https://www.ajc.com/news/state--regional-govt--politics/safety-advocates-georgia-distracted-driving-law-first-step/MwdEUWLf6BfGlTzdL16zkI.

World Health Organization. (2011). *Mobile Phone Use: A Growing Problem of Driver Distraction*. Geneva, Switzerland, World Health Organization. https://www.who.int/publications/i/item/mobile-phone-use-a-growing-problem-of-driver-distraction.

Yager, C. E. 2013. "An Evaluation of the Effectiveness of Voice-to-Text Programs at Reducing Instances of Distracted Driving." Southwest University Transportation Center Report SWUCT/13/600451-00011-1. Contract DTRT12-G-UTC06. College Station: Texas Transportation Institute.

Yager, C. E., J. M. Cooper, and S. T. Chrysler. 2012. "The Effects of Reading and Writing Text-Based Messages While Driving." *Proceedings of the Human Factors and Ergonomics Society, 56th Annual Meeting*, 2196–200.

Yan, W., W. Xiang, S. C. Wong, X. Yan, Y. C. Li, and W. Ho. 2018. "Effects of Hands-Free Cellular Phone Conversational Cognitive Tasks On Driving Stability Based on Driving Simulation Experiment." *Transportation Research Part F* 58: 264–81.

Zhang, B., J. de Winter, S. Varotto, R. Happee, and M. Martens. 2019. "Determinants of Take-Over Time from Automated Driving: A Meta-Analysis of 129 Studies." *Transportation Research Part F* 64: 285–307.

Chapter 7

Automated Vehicles and Environmental Justice

Addressing the Challenges Ahead

Shane Epting

Several researchers have criticized transportation systems in the United States because they harm historically marginalized people and vulnerable populations. Often these cases count as acts of environmental injustice. Some instances focus on health impacts, such as increases in respiratory illness due to excessive carbon emissions from automobiles. Other issues concern community displacement that results from highway implementation. For pinpointing culpability in such cases, frameworks that employ insights from environmental justice (EJ) are incredibly beneficial. Using these methods, we can identify how and when automated vehicles (AVs) could help or hinder these populations. This chapter aims to show how cities can appeal to the principles of EJ to incorporate these new technologies into the urban landscape in an ethical manner.

There are numerous reasons to criticize transportation systems in the United States (for example, Bullard and Johnson 1997; Bullard, Johnson, and Torres 2004). We can put these criticisms into categories that allow us to analyze such problems to understand how they can harm different social groups. By employing an EJ framework to assess how marginalized groups suffer from the effects that transportation systems help produce, we can pinpoint precisely how these populations deal with such unfortunate realities. In turn, we can use these findings to deliver better transportation systems. While such an agenda represents one goal that connects EJ to transportation, it must contend with complications accompanying emerging transportation technologies.

For instance, researchers and industry leaders predict that the future is arriving soon with AVs (Higgins 2019). Introducing AVs into urban settings could improve or complicate existing transportation systems, which will require classifying different kinds of EJ issues for AVs. One benefit of thinking about advancing transportation systems by making room for AVs is that these technologies can improve the conditions that lead to the harmful outcomes that stem from many transportation systems—if the predictions about AVs come true (Epting 2019a). To see the possibilities these technologies could create, we cannot analyze them in a vacuum, paying attention to their inherent qualities. Those activities are great for the armchair. Yet we need to think about them as they would appear on the street. That notion underscores the purpose of this chapter.

Although the previously mentioned powers of prediction remain underdeveloped, researchers of all stripes can continue to investigate the likelihood of such realities. They can develop questions that can help us determine how we want to guide future technologies for the common good, as these technologies will likely affect the urban condition. The questions for EJ must pay attention to how AVs can or cannot mitigate such harm, along with the degree that the people who suffer are part of the discussions about alleviating injuries.

Understanding the complexities of implementing AVs into existing transportation systems that carefully consider EJ is this chapter's goal. In turn, it looks at two broad aspects that pertain to marginalized groups that require an examination to avoid harm (and/or dismantle oppression). The first is easing the burden that those current systems of transportation cause. The second is how to do it. When considered together, these concerns represent two far-reaching yet distinct challenges that transportation specialists must address when thinking about AVs and EJ.

Additionally, there are numerous obstacles that transportation professionals must overcome due to the specific manner of how each city is situated, historically and geographically (Bullard and Johnson 1997). To help address such concerns, illustrating the broad contours of bringing AVs into transportation systems will provide a launchpad for approaching specific instances of environmental injustice. In turn, transportation professionals and researchers will have a starting point that helps them identify considerations that, if taken into account, should help them develop solutions that apply to unique cases in transportation history that were not mindful of EJ. We should care about this topic because this history has numerous examples of environmental harm. If we want a future for AVs that do not resemble the past, we must plan wisely and accordingly.

Undertaking this task means that we must examine how an EJ framework can benefit the work of transportation professionals. This idea underscores why the next section of this chapter looks at the particular model of EJ that will underpin our approach: Robert Figueroa's (2006) EJ paradigm, which draws from distributive justice and political recognition.[1] Once we understand why this approach can assist transportation professionals and researchers, I illustrate how several transportation problems in the United States would benefit from applying this framework. The motivation behind using this approach is to pave a pathway forward, away from harm, toward an environmentally just future with AVs. However, one could criticize this approach as being overly optimistic. In closing, I address these concerns by examining some of the possible real-world challenges that will require additional study.

SETTING THE STAGE WITH ENVIRONMENTAL JUSTICE

The basic idea behind EJ is to protect marginalized groups from environmental harm that is unevenly spread across society (Schlosberg 2007). Early issues in this area dealt with concerns about minority populations and toxic waste, pollution, and illegal dumping, focusing on health disparities (Bullard 2005). Yet as attention to such topics expanded, there was a corollary movement in the academic sphere of EJ studies (Bullard 2005). Efforts increased regarding how to conceptualize the scope and limits of EJ. Since then, there have been many scholars who consider what should go into an EJ framework (Schlosberg 2007).

While arguing in the literature about EJ frameworks, Robert Figueroa (2006) proposed what he calls the EJ paradigm. His approach has broad appeal due to its pluralistic nature that, as mentioned earlier, uses aspects of distributive justice and political recognition. This approach sets him apart from other researchers working in EJ studies. For instance, David Schlosberg (2007) notes that Figueroa is the only working philosopher who makes identity and recognition central dimensions of his approach to EJ. In turn, Figueroa (2006) maintains that we are confronting an issue that falls within the scope of environmental injustice whenever a marginalized or vulnerable group suffers any number of harms so that the mainstream population can benefit.

Such harms involve elements of distributive justice, such as detriments to physical health (for example, asthma). They also cover damages to mental health (for example, stress) and less evident considerations such as harms to culture. For instance, an example of this kind of damage is when a neighborhood is destroyed to expand roadways (Payne 2015). These harms require additional fleshing out to understand why these areas deserve inclusion into paradigms of EJ. In addition to these elements that fall within the umbrella of distributive justice, Figueroa (2006) also champions recognition, making it a primary tenet of his EJ paradigm. This notion for him entails that people who are affected by policy decisions should have the ability to weigh in on them. Through creating this multidimensional approach to EJ, Figueroa lays the groundwork for us to approach issues such as transportation systems, which, in the long run, will help us determine ways to incorporate AVs into our cities justly. Consider, for instance, that in addition to the harms mentioned earlier, a party can be responsible for damages that affect a marginalized group's culture, history, and established forms of knowledge (Figueroa 2006). Although these dimensions are challenging to include in empirically based reports, there are many instances wherein cultures have been wronged, counting as acts of environmental injustice.

For instance, when it comes to transportation systems, researchers have illustrated several cases where decisions have been made resulting in the displacement and near destruction of minority communities. In Portland, Oregon, the Alberta neighborhood, a traditionally African American community, was displaced due to rising rent costs associated with new light rail lines and services (Scott 2012). In Dallas, Texas, planners put a highway right through a historically African American neighborhood, a case that local historians investigate actively (Payne 2015; Welfen 2015). Alisha Volante (2015) provides a thorough documentation of how the Rondo community in St. Paul, Minnesota, also had to deal with such conditions. Volante (2015, 2) makes this point evident:

> Historically, St. Paul's Rondo neighborhood drew in many African Americans migrating north from southern states throughout the twentieth century. Prior to the 1956 construction of highway 94 the Rondo neighborhood held a burgeoning working class community connected by men and women's social clubs, many religious denominations, and community centers like Hallie Q. Brown. The construction of Highway 94 cut Rondo's neighborhoods in half, an act that threatened to dissolve the community; families were forced to move from their homes, entrepreneurs forced to shut down businesses, and community centers forced to relocate.

In these incidents, these cases qualify as acts of environmental injustice because the smaller, historically marginalized population suffered injuries to their cultures while the goals of the majority population advanced (Figueroa 2006). There is also the matter of harm to the family and personal time of minority communities. For instance, it has been shown that minority populations and immigrant groups often depend on public transportation, spending considerable time and money riding public transport systems that are quite taxing of time and pocketbook (OPAL Environmental Justice Portland 2012).

In turn, these kinds of actions fit the description of harm that Figueroa outlines in the EJ paradigm.

While the examples mentioned here reveal the less obvious kinds of cases, there are also the more familiar instances that concern matters for the health of marginalized populations, aspects that most EJ frameworks criticize (Schlosberg 2007). These cases include but are not limited to asthma and respiratory illnesses associated with living continuously near busy highways (McEntee and Ogneva-Himmelberger 2008). These cases are well known and discussed in the literature. For example, these issues have troubled marginalized populations in inner-city Atlanta (Lazarus 2001) and ones in the Bronx in New York City (Maantay 2007).

Despite having such a significant impact on the lives of marginalized groups, one can argue that these populations lack the meaningful ability to participate in the decisions that led to the harms that they endured. In such cases, the damages were twofold. First, there were the harms associated with culture and health. Second, there were also the harms of not being recognized politically. The systematic denial of not having a voice in the policy decision led to the actions that caused the damages mentioned earlier.

While the cases here are unfortunate, they illustrate that restorative or reparative measures are required in some instances. The hope is that transport and urban planners can take these lessons to heart, avoiding the mistakes of transportation's past when thinking about the "best (moral) practices" that require embracing actions that can set the course straight. To illustrate a more environmentally just way forward for the future of AVs, I explore some of the appeals that these technologies have for such a task, identify some potential pitfalls, and provides some suggestive items for planners and engineers to consider in the following section.

ENVIRONMENTAL JUSTICE AND AUTOMATED VEHICLES: LOOKING BACKWARD

The inherent appeal of Figueroa's EJ paradigm is that it covers many of the areas of concern that are outlined earlier for dealing with the ways that transportation systems impact marginalized groups. If we consider that AVs can relieve such burdens, then we should think about them as technologies that can mitigate the harm that current systems help create. Thought about in such a manner, an EJ framework can serve us in a backward-looking fashion. For this task, some of the work is complete, at least in the sense that we have established precedent cases as exemplars, such as the instances mentioned earlier. This notion suggests that when moving forward with implementing AVs, transport professionals need to view these technologies as efforts to relieve these harms and, at the same time, plan for them in a way that does not exacerbate those conditions—or create worse problems.

To understand the magnitude of the promise of AVs, let us turn to some of the predictions made. In turn, we will see some of the potentials that these vehicles hold. For example, one prediction is that they will reduce auto emissions, which can improve air quality (Brown, Gonder, and Repac 2014; Iglińskia and Babiak 2017). Research shows that harmful air pollution in metropolitan areas comes from gasoline-powered vehicles that congest roadways (Stone, Santoni de Sio, and Vermaas 2019). Suppose AVs reduce traffic flow and are developed as electric vehicles. In that case, they could contribute to efforts to reduce pollution that harms people, such as the populations in Atlanta and Queens mentioned earlier and in similar metropolitan environments that suffer from similar conditions (Epting 2019a).

Along with this consideration, there is the idea that AVs will be significantly efficient compared to traditional vehicles (for example, Watzenig and Horn 2017). This notion suggests that they will make more ethical use of our highways, boulevards, and streets. By making better use of our cities' transportation infrastructure, one could argue that there could be less of a need to expand it in traditional ways such as widening highways. If this prediction comes true, then it seems less likely that neighborhoods will be subject to the conditions wherein marginalized people are displaced (again, in some cases), premised on the idea that adding additional highway lanes will benefit the common good. Another benefit from significantly increasing the efficiency of cities' mobility networks is that people who depend on transport services would see the duration of their travels on transit decrease, providing them with more time to spend with their families. Considering that time has emerged as a significant consideration for urban mobility, this notion suggests that advanced study is required to uncover the complexities of such situations (Nordström, Hansson, and Hugosson 2019).

If AVs can achieve these feats, they could come about as unintended consequences. Assuming this turns out to be the case, then people suffering from the ill effects of transportation systems will benefit. However, this point reflects poorly on the transportation professionals tasked with creating mobility systems that produce benefits instead of harm. It shows that one of the possible ways that marginalized groups could have their situation improved is through unintentional means. This point suggests the obvious: the primary motivation behind AV technology is not to ease the burdens of the people who need a solution to their transportation problems, but instead the task is to sell AVs to people who can afford them. In turn, a future with AVs would eventually benefit people who are suffering in the long term. This version of "trickle-down transportation planning" rests on several assumptions.

First, it is consequentialist, which means that if the prophecies of AVs do not come true, then there will be nothing to trickle down. If the "AV utopia" version of reality does not manifest, then everyone loses. Yet the people who must endure environmental hardships must continue to suffer from a transportation network that, one could argue, remains largely unsympathetic to their struggles. The latter portion of the statement implies that people who need help the most depend on professionals who do not have their interests at the forefront of their thoughts, but even a peripheral consideration might help. Yet the point here is not to vilify transportation professionals, only to indicate where they can find some of the problems that concern their occupations.

While they should not give up on solving transportation problems of the past that are causing harm in the present, implementing AVs into existing cities does provide them with an opportunity to cover some of this territory.[2] That is, it is unreasonable to hold today's transportation professionals accountable for systems designed decades ago. However, going forward, they can aim to set things right. One way to approach the problem requires us to bring the recognition aspect of Figueroa's approach into view. This idea holds that if we include people who are subject to environmental injustice in the conversation about future transportation policies and AVs that will impact their lives, then we have diverged from the starting point of yesterdays' planners who help create the harmful conditions that resulted in transportation systems that remain environmentally unjust. In turn, we must look at this aspect of an EJ assessment measure as forward looking, a needed addition to the backward-looking perspective examined earlier. In the following section, I discuss the benefit of employing this method.

ENVIRONMENTAL JUSTICE AND AUTOMATED VEHICLES: LOOKING FORWARD

The previous section paid attention to the considerations for distributive justice that transportation specialists need to bear in mind. However, by bringing forward-looking measures that appeal to political recognition into view, they would be adhering to the requirements for EJ, according to Figueroa's framework. Moreover, one could posit that engaging in such practices could prevent additional environmental injustices from occurring. If transportation planners and engineers were to consider these points seriously, then they would have to design for AVs *with* the people who would disproportionally be affected by them in transportation networks.

Yet engaging in this practice might not yield the results that transportation specialists hope they would receive. One could speculate that planners want people to agree with their plans, but their expectations might be too ambitious. For instance, if one is tasked with planning for a future wherein AVs dominate the cityscape, anything that goes against this idea could be seen as a hindrance. For example, people whose lives are based on public transit might want to improve those systems rather than focus on AVs. If this view turns out to be the case, then it would require additional consideration. People who enjoy the sense of "mobile community" inherently to bus ridership would have this aspect of their lives stripped from them—so only a future with AVs serving as the primary form of highly efficient transportation. In turn, bringing such concerns about AVs into view would be counterproductive. The most efficient course of action for paternally minded transportation specialists would mean that marginalized voices should remain silent. Yet if they want to help create a just future for AVs, they should pay attention to the recognition aspect of EJ.

Exercising this option, however, could put a future for AVs at risk of severe delay. For instance, if marginalized groups would prefer to have current systems operate in a manner that relieves harm today, then giving their insights serious weight could entail deploying AVs onto city streets could come with additional setbacks. This view also runs the risk that these groups have no interest in AVs instead of wanting a future with efficient public transportation, as mentioned earlier. If this were the case, it would create an irreconcilable tension between planners and marginalized groups, which could call for a rethinking of a future with AVs as the primary mode of urban mobility, unless transportation specialists wish to embrace the attitudes of transportation planning's history mentioned earlier.

To avoid this scenario, one could argue that additional research areas will require significant attention through advanced study. This point suggests that marginalized groups could prevent a future for AVs from materializing, meaning that AV advocates and transportation specialists must bend to their desire. However, working together meaningfully could create environmentally just outcomes. Gaining this understanding requires that we give the areas in the next section careful consideration.

FUTURE AREAS OF RESEARCH

Bearing in mind that this chapter has identified two primary areas of concern for AVs and EJ, future research should focus on how AVs can reduce or eliminate environmental harm and promote meaningful participation in the decisions that will shape AV policy. Regarding the former, planners and engineers who are researching the myriad ways that AVs could benefit the urban condition are well positioned to discover ways that these technologies could improve mobility networks to bolster other efforts to promote an environmentally just society. In terms of the latter, social scientists and urban planners could focus on ways to develop mechanisms to improve how marginalized groups can meaningfully participate in the design of transportation systems, which would include how AVs are introduced into society.

The issues covered in this chapter only address AVs from an EJ perspective. Yet several other groups and issues also require attention. For instance, Samantha Noll and Laci Hubbard-Mattix (2019) exhibit how thinking about transportation issues through an intersectional lens can benefit how we think about gender inequality; this also demands advanced study. Noll and Hubbard-Mattix's approach suggests that the future of transportation, especially AVs, will require significant attention from researchers across the academy.

Bearing in mind that if the predictions of AVs come true, we will need to assess how these vehicles could affect other stakeholder groups and entities beyond the public, including ecosystems and nonhuman animals, along with urban artifacts such as infrastructure and historic neighborhoods (Epting 2019b). Due to this condition, turning to an emerging research area referred to as "complex moral assessments" might help us think through broader issues in transportation justice (Epting 2019b). Applying the lessons from EJ and AV could make a needed contribution to such an endeavor.

CONCLUSION

Although we cannot fully grasp how AVs will affect society, scholars hold that they could completely reshape it (Fraedrich et al. 2015; Bagloee et al. 2016; Maurer et al. 2016). Such a notion entails that researchers dedicated to examining EJ issues in society must keep watch, ensuring that these concerns receive adequate attention. This chapter shows that there are two kinds of challenges when it comes to making room for AVs in existing transportation systems: backward looking and forward looking. The former appeal to principles of distributive justice, and the latter brings consideration for political recognition into view.

Due to these two aspects, employing a pluralistic EJ framework revealed two distinct challenges for transportation specialists. The first requires them to mitigate existing harm, which could be feasible—if the predictions about AVs' numerous benefits come true (Epting 2019a). However, if they fail to achieve such goals, planners and engineers wasted time, money, and resources, at least when viewed through a lens of EJ. To make matters worse, existing issues wherein marginalized groups are suffering will not receive attention. New problems that AVs help cause could also emerge, forcing us to deal with more problems than anticipated.

In terms of recognition, this chapter posits that respecting marginalized people requires including their views in the decision-making process and could mean shelving one's ambitions for AVs until the current problems with a given transportation system are solved. While this condition could easily fall outside of the established traditions of transportation professions and comfort zones of specialists, the requirements for EJ, as expressed earlier, should force us to deal with these notions. If the future of urban mobility is not to resemble its past, then such efforts deserve consideration to help create those conditions.

NOTES

1. I examine Figueroa's approach more thoroughly in Epting 2015.
2. Dealing with competing interests between preexisting injustices and anticipated future transport needs is a topic that Karel Martens (2016) examines thoroughly.

REFERENCES

Bagloee, Saeed, Madjid Tavana, Mohsen Asadi, and Tracy Oliver. 2016. "Autonomous Vehicles: Challenges, Opportunities, and Future Implications for Transportation Policies." *Journal Modern Transport* 24 (4): 284–303.

Bullard, Robert, and Glenn Johnson, eds. 1997. *Just Transportation: Dismantling Race and Class Barriers to Mobility*. Gabriola Island, BC: New Society Publishers.

Bullard, Robert, Glenn Johnson, and Angel Torres, eds. 2004. *Highway Robbery: Transportation Racism and New Routes to Equity*. Boston: South End Press.

Bullard, Robert. 2005. "Introduction." In *The Quest for Environmental Justice: Human Rights, and the Politics of Pollution*, edited by Robert Bullard. San Francisco: Sierra Club Books.

Brown, Austin, Jeffrey Gonder, and Brittany Repac. 2014. "An Analysis of Possible Energy Impacts of Automated Vehicles." In *Road Vehicle Automation*, edited by Gereon Meyer and Sven Beiker, 137–53. Cham, Switzerland: Springer.

Epting, Shane. 2015. "The Limits of Environmental Remediation Protocols for Environmental Justice Cases: Lessons from Vieques, Puerto Rico." *Contemporary Justice Review* 18 (3): 352–65.

Epting, Shane. 2019a. "Transportation Planning for Automated Vehicles—Or Automated Vehicles for Transportation Planning?" *Essays in Philosophy* 20 (2): 1–17.

Epting, Shane. 2019b. "Automated Vehicles and Transportation Justice." *Philosophy & Technology* 32 (3): 389–403.

Figueroa, Robert. 2006. "Evaluating Environmental Justice Claims." In *Forging Environmentalism: Justice, Livelihood, and Contested Environments*, edited by Joanne Bauer, 360–61. Armonk, NY: M. E. Sharpe.

Fraedrich, Eva, Sven Beiker, and Barbara Lenz. 2015. "Transition Pathways to Fully Automated Driving and Its Implications for the Sociotechnical System of Automobility." *European Journal of Futures Research* 3 (1): 1–11.

Higgins, Tim. 2019. "Driverless Cars Tap the Brakes After Years of Hype; Developers Take a More Cautious, Low-Key Approach in Testing and Talking About Autonomous Vehicles After Uber Crash." *The Wall Street Journal*. https://www.wsj.com/articles/driverless-cars-tap-the-brakes-after-years-of-hype11547737205.

Iglińskia, Hubert, and Maciej Babiak. 2017. "Analysis of the Potential of Autonomous Vehicles in Reducing the Emissions of Greenhouse Gases in Road Transport." *Procedia Engineering* 192: 353–58.

Lazarus, Richard. 2001. "Highways and Bi-Ways for Environmental Justice." *Cumberland Law Review* 31 (3): 569–97.

Maantay, Juliana. 2007. "Asthma and Air Pollution in the Bronx: Methodological and Data Considerations in Using GIS for Environmental Justice and Health Research." *Health and Place* 13 (1): 32–56.

Martens, Karel. 2016. *Transport Justice: Designing Fair Transportation Systems*. Abingdon: Routledge.

Maurer, Markus, J. Christian Gerdes, Barbara Lenz, and Hermann Winner, eds. 2016. *Autonomous Driving: Technical, Legal, and Social Aspects*. Springer.

McEntee, Jesse, and Yelena Ogneva-Himmelberger. 2008. "Diesel Particulate Matter, Lung Cancer, and Asthma Incidences along Major Traffic Corridors in MA, USA: A GIS Analysis." *Health & Place* 14 (4): 817–28.

Noll, Samantha, and Laci Hubbard-Mattix. 2019. "Health Justice in the City: Why an Intersectional Analysis of Transportation Matters for Bioethics." *Essays in Philosophy* 20 (2): 1–16.

Nordström, Maria, Sven Ove Hansson, and Muriel Beser Hugosson. 2019. "Let Me Save You Some Time . . . On Valuing Travelers' Time in Urban Transportation." *Essays in Philosophy* 20 (2): 206–29.

OPAL Environmental Justice Portland. 2012. *Adela's Journey*. https://www.youtube.com/watch?v=IjxebYmey44.

Payne, Briana. 2015. *Oral History of Bonton and Ideal Neighborhoods in Dallas, Texas*. Denton, Texas. UNT Digital Library. http://digital.library.unt.edu/ark:/67531/metadc848166/.

Schlosberg, David. 2007. *Defining Environmental Justice: Theories, Movements, and Nature*. New York: Oxford University Press.

Scott, Aaron. 2012. "By the Grace of God." *Portland Monthly*. https://www.pdxmonthly.com/articles/2012/2/17/african-american-churches-north-portland-march-2012.

Stone, Taylor, Filippo Santoni de Sio, and Pieter Vermaas. 2019. "Driving in the Dark: Designing Autonomous Vehicles for Reducing Light Pollution." *Science and Engineering Ethics*, 1–17. https://doi.org/10.1007/s11948-019-00101-7.

Volante, Alisha. 2015. *The Rondo Neighborhood & African American History in St. Paul, MN: 1900s to Current*. University of Minnesota. https://conservancy.umn.edu/handle/11299/178547.

Watzenig, Daniel, and Martin Horn, eds. 2017. *Automated Driving: Safer and More Efficient Future Driving*. Cham, Switzerland: Springer International Publishing.

Welfen, Craig. 2015. *Bonton + Ideal: A Dallas Neighborhood Stories Film* [Film]. Vancouver: BC Workshop.

Chapter 8

Vehicles of Change

The Role of Public Sector Regulation on the Road to Autonomous Vehicles

Patrick Schmidt and Jeremy Carp

Technology's inventors have never been shy about their ambitions. When modernism as a project appeared to be collapsing in the 1960s, faith in the future was rekindled by those who saw liberatory potential in new inventions (Turner 2006; Bartlett 2017). Grand claims have been made across sectors, as people have looked to developments in information technology (Deb 2014), nanotechnology (Mangematin and Walsh 2012), and biotechnology (Evens and Kaitin 2015), among others. The "future shock" accompanying this rush of innovation stems in part from the unprecedented rate at which such technologies have evolved and the scope of such changes across many facets of human activity (Allenby 2013; McGrath 2013). Joining this cadre of potentially transformative technologies, an outpouring of commentary has seen autonomous vehicles in a revolutionary light, promising a drastic reduction in traffic deaths, unprecedented mobility for the elderly and disabled, a massive change in urban planning as parking and roads are refashioned, and environmental benefits from the efficient use of vehicle fleets (Fagnant and Kockelman 2013; NHTSA 2016b; Anderson et al. 2016). It has undoubtedly helped that this outpouring of interest in autonomous vehicles has coincided with the desire to see environmental sustainability at the heart of social change (Garforth 2018).

What are the barriers to reaching these visions? To be sure, one of the constraints on techno-optimism are the problems with the technologies themselves. Driving is a complicated human activity, dependent not only on the vehicle's capabilities but also innumerable changing conditions from weather to pedestrians ("Driverless Cars Are Taking Longer" 2019). However steady and impressive the progress, some technologies still fall shy of expectations. Nor are consumers convinced that they want to give up driving to a computer. The widespread adoption of autonomous vehicles is far from guaranteed. Still, vehicles with varying degrees of autonomous capabilities are on the roads today, and although our sights may be shifting toward a longer-term horizon, many commentators remain impressed with the rate of change.

Rather than examine the technological or business propositions surrounding driverless cars, in this chapter we reflect on the challenges facing public sector regulation as a midwife of technosocial transition. Automobiles and transportation are heavily regulated in all aspects of their operation and their use. There are countless nodes where regulators—of which there are many—could use their authority in order to block, stall, shape, or facilitate this emerging technology. What should regulators do, if anything, to facilitate the positive potential of the technology? That question implicates the descriptive and explanatory questions at the heart of an institutional response to rapidly changing technology: How have regulators responded to date, and what does this tell us about the objectives and even the capacity of regulators to respond?

Central to our inquiry lies the "pacing problem": when technological change outpaces the ability of institutions to respond in an efficient and effective manner (Marchant 2011a). We resist the tendency to assume that a "lag" between technology and law is always socially counter-productive, like friction gumming the engine driving a new social and economic future. Rather, starting from the recognition that new technologies may also carry heavy risks (World Commission 2005), and that there may be human value stored in the status quo that is worth conserving in a slow transition, how can regulators think creatively about when to intervene concerning technologies that are rapidly evolving (Lin 2007; Mandel 2009; Bosso 2013)? Indeed, moving too quickly to support a new technology could create its own, unintended setbacks. So, then, how should regulators weigh the costs and benefits of changing with the costs and benefits of not changing, when multiplied by the problem of knowing when to act?

Working within the context of US federal and state regulation, this chapter divides into two sections the problem of how to regulate new technologies for which some speculate there are great, even utopian possibilities. In the first section, we describe some major tensions at play in the response to autonomous vehicles, including the barriers to "getting it right." In the second section, we examine more generally a central problem of responding to rapidly evolving technologies—namely, how to balance the pursuit of the future benefits with the interests in the present. The term "interests" should not disparage the values of those invested in the status quo. Rather, to what extent can and should regulation attempt to modulate the pace of change so that it moves to secure long-term benefits while balancing and minimizing the disruptive effect on people today? Our account emphasizes the institutional and legal environments that constrain those decisions. Although regulation can help to facilitate the commercial success of emerging technologies (Downes 2016) as well as manage their potential risks, it also has drawbacks. We argue that law and technology should be considered coevolutionary partners, such that even if technology itself appears to be an exogeneous force—where innovation is widely distributed and a "shock" sometimes beyond state control—law should be a site for reflective consideration about the ways that technologies are absorbed into the fabric of the community.

THE PERILOUS BRIDGE TO THE FUTURE

Although law and policy are often inextricably involved in conditioning and constructing periods of technological change, it remains helpful to conceptualize technology as an exogenous shock to the law, especially in the early phases of an innovation. Innovations get their start in one place and then diffuse geographically. Within the federal system of the United States, technologies present opportunities for many actors to respond. It isn't clear that we have yet left the "early phase" of the institutional response to autonomous vehicles, but regulatory activity has been blossoming across jurisdictions and a vast range of issues: insurance rules, legal liability, licensing, driver impairment regulation, vehicle testing, traffic regulations, cybersecurity and privacy, business opportunity, workforce development, land use planning, revenues, infrastructure development, and more. A survey of those directions is beyond the scope of this chapter. Rather, we first call to mind the disjunct between the promise of autonomous vehicle technologies and the concerns that guide a significant component of the institutional reaction.

Possible Futures

The enthusiastic early interest in autonomous vehicles was sure of one thing: a future with autonomous vehicles could see the world transformed (Pearl 2017). As former US Transportation Secretary Anthony Foxx remarked in 2016, "the development of advanced automated vehicle . . . technologies . . . may prove to be the greatest personal transportation revolution since the popularization of the personal automobile nearly a century ago" (NHTSA 2016b, 3). But of course, the magnitude of these benefits is always difficult to predict, and as with many technologies, the second wave of assessments has been more cautious than the first.

Regulatory policymaking involves a calculus of value propositions. We are now well past the point of seeing cost-benefit analysis as a neutral exercise to be performed by government economists. Beyond discussions that focus largely on tradeoffs between two competing interests in the present—say, for example, between coal producers and renewable energy companies—the case of autonomous vehicles most powerfully presents a conflict between tangible benefits in the present and uncertain but potentially massive benefits after a societal transition. What does the future hold?

The most concrete prediction concerns the human and economic toll of traffic accidents. In 2015, there were more than 6.3 million reported vehicle accidents resulting in 2.44 million injured people and over 35,000 deaths in the United States alone (NHTSA 2016a). According to National Highway Traffic Safety Administration (NHTSA), 94 percent of all vehicle crashes are attributable to human error (NHTSA 2017). These sources of error include, but are not limited to, fatigue, distraction, excessive speed, misjudgment, and impaired driving. Vehicles with no human operators are never drunk, distracted, fatigued, or otherwise susceptible to human failings (Katyal 2014; Pearl 2017). Plus, even a 25 percent reduction in accidents might save nearly $100 billion annually (Fagnant and Kockelman 2013).

Beyond the direct effects, the knock-on effects are beyond calculation. The technologies will also open up mobility for older citizens, the disabled, people too young to drive, and others without a driver's license. From the benefits that these new freedoms offer come a series of visions. Once able to operate independently of the owner-driver, automobiles may offer personal transportation as a service in a world in which fewer cars exist as services. Untethered in this way, cars wouldn't need to sit parked outside homes and workplaces and could instead remain in active service. The impact on land use could be profound: city codes that have dedicated numerous requirements to the size of parking lots and the width of streets would be outmoded. Crowded urban areas would be gifted vast tracts—a third of the land area of some US cities—as surface parking disappears (Ben-Joseph 2012; Levinson 2015). Beyond that, even, the privacy and comfort of the car, with none of the distraction of driving, could invite riders to commute differently, perhaps unmaking the urban core and changing patterns of urban affordability (Anderson et al. 2016). All the while, by changing roadway capacity, eliminating accidents, and becoming lighter, the national fleet of vehicles could reduce energy use and emissions, possibly dramatically.

If, as the maxim holds, you cannot create an omelet without cracking some eggs, it begs considering whose eggs are on the menu. The revolutionary potential of autonomous vehicles necessarily creates losers from those whose interests are linked to the status quo. The millions of individuals whose careers are implicated, especially long-haul truckers, taxi cab drivers, and bus drivers, head the list. The arrival of ride-sharing services Uber and Lyft, enabled by digital technologies, has already illustrated the potential disruptions facing such providers. Likewise, while many industrial firms such as automakers could profit in a transition to a new fleet of vehicles, not all are equally well positioned to capitalize on it. The further one goes in imagining cities being redesigned for a radically altered understanding of personal transport, the more one would consider the potential resistance that both capital and labor could offer.

The Safety Reflex

It is generally accepted that, absent specific laws or regulations to the contrary, autonomous vehicles are legal in the United States, and governments have tended to a "hands-off" approach—perhaps to a point of underregulation. When the issue rose to public salience, the vast majority of states refrained from passing wide-ranging autonomous vehicle legislation and the federal government declined to initiate any rulemakings in the area of autonomous vehicle safety and design (NHTSA 2016b). There is, effectively, a blank canvas being painted on by federal and state regulators, with a handful of states leading the way in experimentation[1] and federal regulators moving through the lengthy process of federal rulemaking in a number of areas.

The US federal government has maintained a permissive posture toward autonomous vehicle technology. As autonomous vehicles began to emerge in the public imagination and the media, passenger safety has commanded some attention. This emphasis is telling, on the one hand, about how the public first assessed the risk of a driverless car. On the other hand, it reflects the conservatism by which the regulatory "turf" of a body designed for a past technology was the starting point for the innovative technology. By name and mission, the NHTSA was that turf for analysis and rulemaking. Still, the NHTSA had difficulty settling on a clear course of regulatory action and provided little guidance to industry stakeholders. As articulated in its 2016 policy statement, the NHTSA described a variety of possible tools, including requiring manufacturers to report safety data as they bring autonomous vehicles to market, forcing manufacturers to submit to a premarket approval process for autonomous vehicles, and empowering NHTSA to require that manufacturers take immediate action to mitigate safety risks deemed "imminent hazards" posing "immediate risk of death, personal injury, or significant harm to the public" (NHTSA 2016b, 75). When it did first issue policy guidance in 2016 and 2017, amid a wave of urging from those in industry who wanted clarity, NHTSA outlined nonbinding recommendations for state governments and industry stakeholders and emphasized voluntary guidance on the design, testing, and deployment of autonomous vehicles (NHTSA 2017).

The vast majority of US states reacted to the technological horizon with a wait-and-see approach, while a handful of states and the District of Columbia waded into the field through legislation. States did not alter the process by which they approached vehicle regulation, leaving state agencies to establish and revise the rules governing autonomous vehicles, just as they do with traditional vehicles. The laws of early mover states themselves pulled in different directions, including on the one hand express authorization for the unrestricted operation of autonomous vehicles, and on the other specific design mandates and limits on the operation of autonomous vehicles. Among the small number of states that sought to create regulatory space for driverless cars—such as Michigan, Florida, and Tennessee—any person with a valid driver's license could own an autonomous vehicle and facilitate its operation without having the ability to take remote control of the vehicle or be physically present.

On the other side, some jurisdictions—such as California, Nevada, and the District of Columbia—required that a licensed human driver be present and capable of taking manual control of an autonomous vehicle at all times. This mandate effectively required that every autonomous vehicle be equipped with a steering wheel, accelerator, and brake pedal. Nevada further restricted the testing of autonomous vehicles to preapproved geographic areas and external conditions. The District of Columbia restricted autonomous vehicle operation to individuals who qualified for a special autonomous vehicle license.

The frame of safety risks a bludgeoning effect when translated into a regulatory response to new technologies. Especially where those technologies involve biological processes and environmental impacts, safety can drive a version of the precautionary principle, by which caution governs whenever the ultimate effects are unknown or disputed. The guidance that followed from the US Department of Transportation in subsequent years maintained a role for safety but began to introduce balancing priorities. In October 2018, a federal report offered a major complement to safety with a priority of remaining "technology neutral" (NHTSA 2018). The reasoning appears to reflect skepticism about governmental expertise when making choices about technologies. Instead the report made markets the locus for decisions about the best way to serve the public interest. The Department cited "the dynamic and rapid development of automated vehicles" as the leading factor for preferring "competition and innovation as a means of [achieving] safety, mobility, and economic goals" (NTHSA 2018, iv).

Through iterative policy vision statements, then, the US Department of Transportation found its way to a creative tension between "three core interests," themselves multifaceted (NHTSA 2020, 1). The first, "protecting users and consumers," included a first priority of safety, while also including the desire to enhance mobility and accessibility—a key virtue of coming technologies. Simultaneously, the second core interest of promoting "efficient markets" delegates and naturalizes the complex choices that otherwise would fall to regulators who, while well positioned to understand the public benefit in the widespread adoption of autonomous vehicle technology, can also be wary of offending core constituencies—or the elected representatives who defend them. Thus, the stated desire to "remain technology neutral" puts a technology before a different decision maker (NHTSA 2020, 1). If this tension weren't enough, the third core interest—and a serious challenge—has itself been to "facilitate coordinated efforts" through a consistent federal approach.

Notably, regulators have offered evolutionary rather than revolutionary approaches to vehicle safety and performance standards. One touchstone of conservatism in periods of rapid technological change is deference to past institutions themselves: autonomous vehicles are seen as fitting within existing regulatory frameworks, and decision making thus is cabined by priorities of the institution that maintains that framework.

Barriers to Transition

The turn toward safety need not define the whole response to a new technology, of course, and may just offer a necessary pause as the technology emerges. Beyond the "pause," however, there are good reasons to think that the institutional apparatus of law and regulation creates lasting barriers to securing the possible benefits of rapidly evolving technologies. A growing body of literature acknowledges that law and emerging technologies are often poorly aligned, in what one commentator labels the "pacing problem" (Marchant 2011a). No institution of governance was designed to respond efficiently and effectively to fast-moving innovations.

We are long past the point where we would expect the legislative process, whether in the US Congress or state legislatures, to be the primary vehicle for rapid or frequent policy change. While able to delegate regulatory authority to administrative bodies, both in design and practice, legislative decision making is slow and easily bogged down by constituencies' passions and desires. Bicameralism, present in most US legislative bodies, and elaborate rules intentionally constrain efficiency. The partisanship of the present era only compounds the reactive nature of legislative action. Congress and state legislatures are often faced with more issues than time or resources allow them to address (Marchant et al. 2009). As John Kingdon famously argued, policy issues are unlikely to receive attention outside of brief "policy windows" when political feasibility, social urgency, and mature policy solutions combine to allow for legislative action (Kingdon 1995). Although the combination of these factors permits new laws to be enacted or old laws to be adapted during an open window, it may be years before political conditions allow lawmakers to revisit the same issue during a new window.

In the modern era, administrative agencies have wielded greater flexibility to act, though to be sure, US regulatory policymaking is notoriously complex and fraught with delays (Sunstein 1995; Yackee and Yackee 2012). By their nature, regulatory bodies are creations of legislation, however, and so enact the political bargain that drove their creation as much as a century before. The authority agencies possess may be a poor fit for the task of regulating an emerging technology unanticipated decades earlier—including the possibility that the agency with the most expertise may be blocked from regulating the new technology altogether (Moses 2005; Mandel 2009). Such provisions are a form of equity owned by stakeholders in the status quo, who have captured the agency over time (Stigler 1971). In the context of emerging technologies, an incumbent industry benefiting from an existing regulatory scheme or hoping to handicap a new technology may use its clout to prevent an agency from effectively regulating the new technology, or at least from doing so in a manner that necessarily prioritizes the public interest. Similarly, powerful business interests invested in developing new technologies, such as Google or Volvo in the case of autonomous vehicles, may attempt to prevent new regulations designed to promote safety but which impose significant costs or disadvantage specific forms of a technology.

For their part, American courts are similarly ill suited to negotiate society's bargain with promising technologies. Although the common law basis of US courts does suggest adaptability without reliance on legislatures, precedent and analogical reasoning substitutes in its place (Hathaway 2001). Adaptation is possible, then, but the revision or replacement of legal doctrines is slow and piecemeal (Moses 2003; Moses 2007). The litigation process by which this change happens is plodding and focused on individualized disputes. Even when courts see a need to push the law forward, there are strong norms discouraging action, some unequivocal boundaries in their ability to effectively rewrite a statute, and serious questions about the capacity of generalist courts to discern the best courses of action. The faster moving the technology, the worse courts can be. When the developmental path of a technology is highly uncertain, as is the case with autonomous vehicles, "a judicial opinion could be outdated before it is even decided even at the time it is issued" (Marchant 2011a, 24).

PURSUING IDEALS THROUGH IMPERFECT INSTITUTIONS

In the account we have developed, the revolutionary potential of a rapidly evolving technology poses a concomitant problem for governance institutions. The people inhabiting those institutions will make choices that apportion costs and benefits (or opportunity costs) among status quo stakeholders and potential, future beneficiaries. Although these institutions are often staffed by experts and positioned as guardians of the public interest, they also operate with certain inherent limitations: they are susceptible to industry capture, hampered by quickly outdated and at times limited understandings of the technology itself, and poorly positioned to revise policies in need of frequent updates. The inherent tension means that the approaches that are best suited to an institution's capacities are often conservative and designed to frustrate the emergence of the technology; approaches designed to most efficiently promote the sociotechnical transition are furthest from its competence.

Three Paths Considered

In the context of autonomous vehicles, three approaches have already been on display among the various regulatory bodies of the United States: precaution, inaction, and proactivity.

The first, a conservative regulatory response to rapid technological change, has found increasing favor over the past several decades. The "precautionary principle," embodied by the maxim "better safe than sorry" (Wood et al. 2006, 581), holds that the best response is to wait until a technology is found safe or least until an informed regulatory response is available (Sunstein 2005; Light 2017). By restraining the rate at which a new technology can take root and evolve, proponents believe that the precautionary principle provides regulators with additional time and information to design and enact regulatory frameworks, as well as reduces the likelihood that such frameworks will need to be amended based on new information regarding the risks or trajectory of an emerging technology.

In the context of autonomous vehicles, some states have already responded to the emergence of driverless technology by attempting to slow its development. Insisting on the presence of a human driver or features such as a steering wheel, for example, have been small moves that interrupt innovation. A blanket precautionary principle may offer cover to stakeholders who would like to delay the onset of a technology. If that time was then used to develop a plan for economic transition, an "economic precautionary principle" might be justified. Disruption that affects firms and workers, or "retail-level" employees such as truck drivers, may call for massive investment in market transition and worker education. Communities rightly want to avoid spirals of decline, where human misery arises amid new "rust belts" where old technologies once thrived.

The greater fault of a blanket precautionary principle may be the failure to distinguish between different types of technologies. Unlike some forms of biotechnology or the use of nuclear materials, autonomous vehicles present fewer fears of "runaway" technology or unstoppable alteration to the environment. Moreover, the safety question might not be especially convincing when thinking about the past century of automotive transportation: the rate of motor vehicle accidents and fatalities is so high that even after allowing for some severe scenarios driven by cybersecurity threats or unanticipated technological faults, we might reject out of hand any delay in introducing a technology that could bring dramatic reductions in fatalities.

The "pacing problem" of a conservative approach is knowing when to loosen the grip of regulation. From inception to complete transition, some "rapid" technological onsets may take decades. Precaution doesn't mean prevention. Institutions more responsive to constituent interests may be especially slow to react when constituencies have undue influence over a regulatory agency or elected officials. One consequence is to incur the opportunity cost of saved lives and enhanced mobility, but the deeper and unknowable consequence may be the diversion of energy and investment in innovation. Where there's an in-rush of interest and experimentation, a delay could have the effect of strangling the nascent industry and channeling both investment and consumers in different directions (Thierer 2016).

In practice, American courts have answered this challenge. The traditional common law rule that land is owned *usque ad coelum*, or infinitely upward, created uncertainty for early aviation because it would have effectively prohibited every flight over private land as a trespass (Moses 2003; Heller 2008). Courts eventually stepped in to circumscribe the doctrine and allow commercial aviation to move forward. For autonomous vehicles, one would expect that courts will provide flexible interpretations for the terms of the 1949 Geneva Convention on Road Traffic, to which the United States is a party, Article 8 of which requires every vehicle to have a "driver" who is "at all times . . . able to control [it]." Similarly, police, prosecutors, and courts have sufficient discretion to get around requirements—such as the requirement that drivers have at least one hand on the steering wheel at all times—when technological development changes what it means to have "control" of a vehicle.

Not everything can be solved this way, however. New technologies can expose "latent ambiguities" in the terms and concepts contained in existing laws, such as the terms "driver" and "operator," and while the ambiguity may provide sufficient room for encompassing computers within the meaning, some ambiguities can be solved only by a legislature. When making investment decisions, individuals and firms may struggle with this ambiguity, unsure of the liabilities they could incur or the ways in which an autonomous vehicle could be lawfully used.

A second response to rapid technological change is to limit or refrain from regulating a new technology (Kirby 2008). After committing sociotechnical transitions to market forces and existing legal frameworks, regulators inject themselves only when greater clarity has emerged (Demissie 2008). Inaction and simple legalization has been a leading regulatory response to autonomous vehicle technology. As described earlier, most states and the federal NHTSA have taken largely passive roles. Inaction has particular appeal to generalist policymakers who are poorly informed about a technology or who lack the capacity to quickly adapt to the changes in the field. But the greater political appeal may be that it allows policymakers to position themselves as neutral parties when new technologies pose significant (even existential) threats to ways of doing business.

But inaction may suffer some of the same faults of the precautionary principle, especially where the promise of transition to a new sociotechnical future depends on the transformation of structures—such as zoning and land use regulations—that inherently lay with the public sector. For autonomous vehicles, the promise of a dramatic transition in Americans' relationship to the built environment cannot be achieved without affirmative steps taken to promote development around autonomous vehicles. Those policies are highly facilitative; they may alter the direction of investment and technological innovation.

Regulatory inaction can also undermine public confidence in the safety of a new technology. The importance of public confidence cannot be overstated (Sylvester 2009; Obama 2016). If consumers are unwilling to purchase or use a new technology due to concerns about safety—whether justified or not—then the potential benefits of a new technology are virtually guaranteed to remain unrealized (Lin 2007). Importantly, the degree to which a technology is regulated relates to public confidence in at least two ways. First, to the extent that imposing minimum safety standards on a new technology actually reduces the probability of harmful events, the regulated technology is likely to be perceived as safer (Wansley 2016). Because most people first hear about a new technology through the media, highly publicized incidents involving the technology could cause people to overestimate its risks (Sunstein 2002). Second, a growing body of research suggests that public confidence in unfamiliar technologies depends at least in part on their level of regulation (Gollier and Treich 2003; Moses 2005; Marchant et al. 2009). This is especially true in the wake of high-profile accidents involving such technologies. Technologies from genetically modified food to nuclear power have floundered when public confidence was lost; leaving the market to the rationality of consumers' risk assessments may not be the wisest strategy.

A third response to rapid technological change is to enact future-facing regulations that attempt to anticipate or otherwise shape the development of an emerging technology. Proponents of this response acknowledge the limitations of regulatory institutions but view regulation as an important element in the success and safety of new innovations. They maintain that because regulatory institutions struggle to react and adapt to rapid changes in emerging technologies, regulators must design frameworks that anticipate, or attempt to guide, their ongoing and future development (Birnhack 2013; Greenberg 2016). In particular, the approach assumes that regulators must minimize the probability that regulations enacted in response to a new technology will need to be revised or revisited in the future. Although there is no uniform theory of how to avoid such revisions, the two most prominent approaches are (1) to mandate specific characteristics or forms of a technology and (2) to draft technology-neutral laws that focus on achieving a particular state of the world rather than a particular state of a technology (Pidot 2015; Greenberg 2016).

The inherent hazard of the first approach attaches to any sort of central planning and the chance (or likelihood even) that a public sector bet on the technology will create serious distortions in the marketplace. Efforts to spur the development of renewable energy offer one recent example. "It is an unhappy fact of life," Grant Gilmore famously wrote, "that, while we can know the past only imperfectly, we know the future not at all" (Gilmore 1967, 467). The consequences of wrongly forecasting the evolution of a new technology, or attempting to shape its development based on incomplete information or faulty assumptions, can be significant. A future-facing regulation, for example, can lock developers and the marketplace into one suboptimal version of a technology or freeze into law a set of expectations imposed on the technology from a less expert vantage point.

Industry may appreciate incentives to investment, but the field as a whole may hold a greater reliance interest in the stability of public support than democratic winds can provide. Courts, and to a lesser extent agencies, are thought to have a greater capacity than legislatures to meet the reliance interests of private parties. At the same time, the energy of government backing tends to implicate individual liberties. In order to secure or maximize the advantages of driverless cars, government power may be necessary to transition everyone to the use of autonomous vehicles and take human drivers off the road. At least as an ethical intuition, it would be easier for an American government to ban a future technology that no one uses than to ban the traditional technology to which many people have a deep attachment, even if the freedom to self-drive is the freedom to subject yourself to considerably higher risks.

For that reason, then, technology-neutral laws have had significant appeal in a wide range of technological contexts over the past fifty years, such as in Congress's development of a technology-neutral copyright framework (Greenberg 2016). In the realm of autonomous vehicles, states have largely avoided such an approach, however. California is the exception, perhaps owing to the recognition of the state's influence on the automotive industry and the different approach the state has taken to federal regulators on a variety of issues (Korosec 2015). Of course, future-facing laws among the states could still frustrate the development of the technology if those regulations varied widely, and ideally some foundations for autonomous vehicles will come from a nationally coordinated approach. Additionally, technology-neutral regulation can still create barriers to the emergence of new technologies because its "broad, open-textured" language provides flexibility at the expense of consistency and clarity.

Engaging Uncertainty

In essence, regulators of autonomous vehicles and other emerging technologies face a maddening catch-22. On the one hand, the less they regulate or the broader they phrase regulations, the more disconnected law may become from its target or the less clarity and certainty regulated entities and consumers will be able to enjoy. On the other hand, the more certainty and confidence regulators attempt to provide to entities and consumers, the more likely it is that regulations will become outdated, generate future tensions, or pigeonhole a developing technology. In both cases, if regulators block or slow a technology until its likely developmental path and risks are fully understood, then its benefits—including any life-saving applications—will remain unrealized or diluted. How can we know where the balance lies?

For one, regulation must proceed on an assumption of conditionality, by which regulation becomes an explicitly iterative process, with the first regulations of autonomous vehicles as an initial step rather than the ultimate goal. Such an iterative model of regulation would depend upon federal and state regulatory institutions "designed from the outset to expect, anticipate, and be able to . . . recalibrate [regulations] quickly" in response to rapid technological change and new information about risks, benefits, and the effects of existing rules (Marchant 2011, 202). Part of this approach emphasizes the model of planned adaptive regulation, which would keep autonomous vehicle regulation "yoked to an evolving knowledge base" and avoid the significant social and technological pitfalls of regulatory inaction and prohibition (McCray et al. 2010, 952). Although there is no consensus definition of planned adaptive regulation, a planned adaptive approach is generally characterized by two fundamental attributes: (1) "a prior commitment, planned early in the policy's design, to subject the policy to periodic re-evaluation and potential revision," and (2) "a systemic effort or mechanism planned early in the policy's design, to monitor and synthesize new information for use in the re-evaluations" (International Risk Governance Center 2016, 26). In turn, planned adaptive regulation "requires institutionalization of monitoring-adjustment frameworks that allow incremental policy and decision adjustments," *ex post*, "where performance results can be evaluated and the new information can be fed back into the ongoing regulatory process" (Ruhl 2005, 30). A carefully conceived "framework for altering course, rapidly and frequently if conditions warrant, are thus essential ingredients" of a planned adaptive approach to regulation in any context (Ruhl 2005, 30).

Already decisionmakers in the field of autonomous vehicles have practiced some elements of this model. States and NHTSA have been engaged in a process of issuing initial regulations. The next phase of adaption emphasizes the centrality of good information, such as intensive data collection, and careful assessment by independent parties, after which agencies must reconsider in a spirit that allows for course correction. The substantive starting points of regulation are relevant here, however, given the risk that significant restrictions on the design and operation of autonomous vehicles could prematurely foreclose certain paths for technological development and data collection. Initial regulations need to default to the creation of a relatively permissive legal environment constrained only by minimum safety standards. This calls into question the precautionary principle, along with technology-specific rules. The comparison for harm in the field of autonomous vehicles is a technology already widely in use, so while regulation is an important bulwark against unsafe technologies, evidence of comparative advantage—that which reasonably suggests that autonomous vehicles, in design and practice, are at least as safe as traditional driving technology—would help to establish the grounds for continuing permissiveness (Kalra and Paddock 2016).

An adaptive system puts the primacy on the quality and depth of information gathering and assessment. After the emergence of COVID-19 in 2020, much attention turned to the efforts by competing groups (across national lines) to develop and test potential vaccines for safety and efficacy. In the face of that competition, against a background of financial gain and human need, the key regulatory question was the resilience of systems that sought to oversee and hold accountable manufacturers of vaccine candidates. Any decision to change course—to change the marketplace—must be empirically driven and based on rigorous assessments of costs, benefits, and risks. The feedback loops between regulation and revision, and between regulators and stakeholders, depend on the quality of mechanisms for study and accountability. Given the private interests at stake, and the potential for interests to affect these assessments, the development of *a priori* principles, objectives, and standards establishes the currency for credible judgment (McCray et al. 2010). Yet in an area of rapidly changing technology, there is cause for skepticism that this process can occur solely within the folds of expertise-based government agency. Rapid change in the field means that stakeholders among the regulated industry must be incorporated into the process, so that their field-specific knowledge becomes part of a process of continuing, up-to-date monitoring. Nevertheless, separation between the process of information analysis and final decision making remains essential to adaptive systems. Inaction on the part of government sometimes leads to a model of industry self-regulation, which has long been subject to justified skepticism.

Some aspects of adaptive regulation seem intuitive. To an individual, the hallmarks of reflective, even meditative living are time honored. State and federal institutions have struggled to build wisdom and responsiveness into official processes. The procedures governing agency rulemaking, and the long-established cultures among government officials and stakeholders, make it difficult or impossible to adapt to changing information. As one commentator notes, agencies "have not often been rewarded for flexibility, openness, and their willingness to experiment, monitor, and adapt" (Grumbine 1977, 45). As such, for adaptive regulation of autonomous vehicles to succeed, "legislatures must empower [administrative agencies] to do it, interest groups must let them do it, and the courts must resist the temptation to second-guess when [agencies] do in fact do it" (Ruhl 2005, 31). It would, in short, require "substantial change" in existing structures and assumptions underpinning administrative law (Ruhl 2005, 31). This paradigm shift is attainable but will not occur naturally.

A more self-consciously adaptive approach would chart a middle course. Although regulators may determine that autonomous vehicles are best governed with a light touch, the potential safety risks associated with the technology and fallout from highly visible accidents make it critical that some form of government regulation guarantee minimum levels of safety and, just as importantly, sow consumer confidence in the technology. Yet slowing the development of autonomous vehicles in response to potential risks would be equally crude and forgo significant benefits due to comparably small or unproven costs. On the flip side, adaptive regulation avoids the pitfalls of static mandates by responding to, rather than locking in or implicitly anticipating, specific evolutions in autonomous vehicle technology. Instead of making an educated guess and going "all in" with a revolutionary technology, an adaptive approach allows for adjustment based on data and debate. There is no need to "future proof" laws where laws can be revisited more actively.

The costs of this approach, both to agencies and industry stakeholders, are significant. Beyond the costs of information gathering, each revision opens new windows to the influence of interests, whether of the status quo or stakeholders in the future benefits. Government agencies already experience significant risk of regulatory capture (Whitehead 2012). There is also a chance, as mentioned previously, that agency officials would be reluctant to implement such a system due to institutional inertia, a preference for the status quo, or concerns about agency credibility.

Cases such as autonomous vehicles, if nothing else, provide an opportunity for the practitioners of law and policy to engage with their existential limits. The scholarly and professional discourse in an area such as this displays the common assumption that it is possible, through all of the regulatory proposals that get floated and considered, to identify *ex ante* an optimal substantive choice of rules. If through all of this the cumulative response appears to veer toward "inaction," we might start by offering applause. Inaction is, in some cases, simply a way of buying time to draft an optimal regulation, but we might also understand it as a concession of humility and inadequacy. The scale of costs and possible benefits for autonomous vehicles is, arguably, not symmetrical: there are risks, to be sure, but possible gains that could be many magnitudes greater. And, further, the status quo is not inert on the technology: this is not a wholly new invention but one (vehicles) that already has great costs. The regulatory equivalent of the Hippocratic Oath would urge a "do no harm" that leans in favor of action. Appropriately, then, few jurisdictions have established barriers to development.

The focus on the straightforward question of how to respond to the regulatory question at any given time—whether to frustrate, boost, or simply stay out of the way of a new invention—gives way to consideration of the underlying model and process one is following, and to invite a reconsideration of that process. There are numerous barriers to such meta-thinking within the regulatory process. First among these might be the difficulty in thinking self-consciously about the object of regulation: Is this an invention that heralds such a transformation that it deserves a pause in the process? While a critical mass of thinking may emerge around some inventions—say, nanotechnology, 3D printing, and the like—to whose social and ethical ramifications scholars can give serious, sustained attention, inventors are quick to claim utopian possibilities for many creations. Then, as noted earlier, because the hallmark of a rapidly changing technology is, of course, its rapidity, the call for a pause in the process begs its own questions: Can we pin down the features of the object we are attempting to regulate? Can we move quickly enough to redesign a decision-making process? Does that call for a pause itself enact a decision in this case?

The vexing nature of these questions suggests that while it may be virtually impossible, early in the life of a rapidly evolving technology, to both identify an optimal regulatory approach and establish an enduring framework, it does urge closer attention to *ex post* analysis of uncertainty itself and the need to see uncertainty as embedded within certain periods of sociotechnical regulation. Such analyses may be particularly vital for a period in which environmental and biological questions are rising to the fore, a period in which the planet and the human species are closely tied to the outcomes of technological progress. While it may be too late to undo the decisions made regarding autonomous vehicles, the example (and others) will help generate the palette of options for future technologies.

CONCLUSION

The rise of autonomous vehicles has yielded a wave of possibilities, some concrete and some utopian. At the same time, something as revolutionary as autonomous vehicle technology poses threats to the interests of numerous stakeholders. Institutions of all types struggle with a fundamental problem of collective action: how to come to decisions when many parties have an interest in the outcome. The interests of future parties are notoriously difficult to weigh, and whether the body in question is a legislature, expert agency, or court, all will find it difficult to keep up with developments in a rapidly changing field of technology. The defects are meaningful, and the deficits that result are many. Yet there are opportunities to enhance public sector decision making in pursuit of a more self-consciously reflective regulatory process. We must start by interrogating the assumptions that often drive our impulses to regulate and then direct more resources toward information collection and public education. Beyond that, a more holistic approach to the costs, benefits, risks, and ethics—of both the status quo and the vision of things to come—could be vital in charting a course to realize the greatest ambitions for technology.

ACKNOWLEDGMENTS

The authors would like to thank Cary Coglianese and Lane Centrella for their contributions to this project, portions of which appeared in an earlier form as "Autonomous Vehicles: Problems and Principles for Future Regulation," *University of Pennsylvania Journal of Law & Public Affairs* 4 (2018): 81–148.

NOTE

1. Leading the way were California, Nevada, Tennessee, Florida, Michigan, and the District of Columbia.

REFERENCES

Adkisson, Samuel. 2018. "System-Level Standards: Driverless Cars and the Future of Regulatory Design." *University of Hawaii Law Review* 40. https://papers.ssrn.com/sol3/papers.cfm?abstract_id=3122393.

Allenby, Braden. 2013. "The Dynamics of Emerging Technology Systems." In *Innovative Governance Models for Emerging Technologies*, edited by Gary E. Marchant et al., 19–43. Northampton: Edward Elgar Publishing.

Anderson, James A., Nidhi Kalra, Karlyn D. Stanley, Paul Sorensen, Constantine Samaras, and Tobi A. Oluwatola. 2016. *Autonomous Vehicle Technologies: A Guide for Policymakers*. Santa Monica, CA: RAND Corporation.

Bartlett, Jaime. 2017. *Radicals Chasing Utopia*. New York: Nation Books.

Ben-Joseph, Eran. 2012. *ReThinking a Lot: The Design and Culture of Parking*. Cambridge, MA: MIT Press.

Bergeson, Lynn L. 2005. "Avoid Mistakes of the Past: Develop Nano Responsibly." *Environmental Law of Florida* 22: 41.

Birnhack, Michael. 2013. "Reverse Engineering Information Privacy Law." *Yale Journal of Law and Technology* 15: 38–39.

Bosso, Christopher. 2013. "The Enduring Embrace: The Regulatory Ancien Régime and Governance of Nanomaterials in the U.S." *Nanotechnology Law and Business* 9: 381–92.

Canellas, Marc. 2020. "Unsafe at Any Level: The U.S. NHTSA's Levels of Automation Are a Liability Automated Vehicles." *Communications of the ACM* 63: 31–34. https://papers.ssrn.com/sol3/papers.cfm?abstract_id=3567225.

Craig, Robin, and J. B. Ruhl. 2014. "Designing Administrative Law for Adaptive Management." *Vanderbilt Law Review* 67: 7.

Deb, Sagarmay. 2014. "Information Technology, Its Impact on Society and Its Future." *Advances in Computing* 4: 25.

Demissie, Hailmichael Teshome. 2008. "Taming Matter for the Welfare of Humanity: Regulating Nanotechnology." In *Regulating Technologies: Legal Futures, Regulatory Frames and Technological Fixes*, edited by Roger Brownsword and Karen Yeung, 327–40. Oxford: Hart Publishing.

Downes, Larry. 2016. "The Right and Wrong Way to Regulate Self-Driving Cars." *Harvard Business Review*, December 6. https://hbr.org/2016/12/the-right-and-wrong-ways-to-regulate-self-driving-cars/.

"Driverless Cars Are Taking Longer Than We Expected. Here's Why." 2019. *New York Times*, July 14, 2019. https://www.nytimes.com/2019/07/14/us/driverless-cars.html.

Evens, Ronald, and Kaitin Kenneth. 2015. "The Evolution of Biotechnology and Its Impact on Health Care." *Health Affairs* 34: 210–19.

Fagnant, Daniel J., and Kara M. Kockelman. 2013. *Preparing a Nation for Autonomous Vehicles*. Washington, DC: Eno Center for Transportation. https://www.enotrans.org/wp-content/uploads/AV-paper.pdf.

Garforth, Lisa. 2018. *Green Utopias: Environmental Hope Before and After Nature*. Cambridge: Polity.

Gilmore, Grant. 1967. "On Statutory Obsolescence." *Colorado Law Review* 39: 461–77.
Gollier, Christian, and Nicholas Treich. 2003. "Decision-Making Under Scientific Uncertainty." *Journal of Risk & Uncertainty* 27: 97.
Greenberg, Brad A. 2016. "Rethinking Technology Neutrality." *Minnesota Law Review* 100: 1495–562.
Grumbine, R. Edward. 1977. "Reflections on 'What Is Ecosystem Management?'" *Conservation Biology* 11: 41–47.
Hathaway, Oona. 2001. "Path Dependence in the Law: The Course and Pattern of Legal Change in a Common Law System." *Iowa Law Review* 86: 601–66.
Heller, Michael. 2008. *The Gridlock Economy: How Too Much Ownership Wrecks Markets, Stops Innovation, and Cost Lives.* New York: Basic Books.
International Risk Governance Center. 2016. *Conference Report: Planning Adaptive Risk Regulation.* https://irgc.org/irgc-international-conference-planning-adaptive-risk-regulation/.
Kalra, Nidhi, and Susan M. Paddock. 2016. *Driving to Safety: How Many Miles of Driving Would It Take to Demonstrate Autonomous Vehicle Reliability?* Santa Monica, CA: RAND Corporation. https://www.rand.org/content/dam/rand/pubs/research_reports/RR1400/RR1478/RAND_RR1478.pdf.
Katyal, Neal. 2014. "Disruptive Technologies and the Law." *Georgetown Law Journal* 102: 1685–89.
Kingdon, John W. 1995. *Agendas, Alternatives, and Public Policies.* London: Longman Publishing.
Kirby, Michael. 2008. "New Frontier: Regulating Technology by Law and 'Code.'" In *Regulating Technologies: Legal Futures, Regulatory Frames and Technological Fixes*, edited by Roger Brownsword and Karen Yeung. New York: Bloomsbury Academic.
Korosec, Kirsten. 2015. "Google Is 'Disappointed' with California's New Self-Driving Car Rules." *Fortune*, December 16. http://fortune.com/2015/12/16/google-california-rules-self-driving-cars/.
Levinson, David. 2015. "Climbing Mount Next: The Effects of Autonomous Vehicles on Society." *Minnesota Journal of Law, Science & Technology* 16: 787–809.
Lin, Albert C. 2007. "Size Matters: Regulating Nanotechnology." *Harvard Environmental Law Review* 31: 349–408.
Light, Sarah E. 2017. "Precautionary Federalism and the Sharing Economy." *Emory Law Journal* 66: 333–94.
Mandel, Gregory N. 2009. "Regulating Emerging Technologies." *Law, Innovation & Technology* 1: 75–92.
Mangematin, Vincent, and Steve Walsh. 2012. "The Future of Nanotechnologies." *Technovation* 32: 157.
Marchant, Gary E. 2011a. "The Growing Gap Between Emerging Technologies and the Law." In *The Growing Gap Between Emerging Technologies and Legal-Ethical Oversight*, edited by Gary E. Marchant, Braden R. Allenby, and Joseph R. Herkert. New York: Springer.

Marchant, Gary E. 2011b. "Addressing the Pacing Problem." In *The Growing Gap Between Emerging Technologies and Legal-Ethical Oversight*, edited by Gary E. Marchant, Braden R. Allenby, and Joseph R. Herkert. New York: Springer.

Marchant, Gary E., Kenneth W. Abbott, and Douglas J. Sylvester. 2009. "What Does the History of Technology Regulation Teach Us About Nano Oversight?" *Journal of Law, Medicine & Ethics* 37: 724–31.

Marchant, Gary E., et al. 2013. *Innovative Governance Models for Emerging Technologies*. Northampton: Edward Elgar Publishing.

McCray, Lawrence E., Kenneth A. Oye, and Arthur C. Petersen. 2010. "Planned Adaptation in Risk Regulation: An Initial Survey of U.S. Environmental, Health, and Safety Regulation." *Technological Forecasting & Social Change* 77: 951–59.

McGrath, Rita. 2013. "The Pace of Technology Adoption Is Speeding Up." *Harvard Business Review.* https://hbr.org/2013/11/the-pace-of-technology-adoption-is-speeding-up.

Moore, Gordon E. 1975. "Progress in Digital Integrated Electronics." *Technical Digest*: 11–13.

Moses, Lyria Bennett. 2003 "Adapting the Law to Technological Change: A Comparison of Common Law and Legislation." *University of New South Wales Law Journal* 26: 394–417.

Moses, Lyria Bennett. 2005. "Understanding Legal Responses to Technological Change: The Example of In Vitro Fertilization." *Minnesota Journal of Law, Science & Technology* 6: 505–618.

Moses, Lyria Bennett. 2007. "Recurring Dilemmas: The Law's Race to Keep Up with Technological Change." *University of Illinois Journal of Law, Technology, and Policy*: 247–69.

National Conference of State Legislatures. 2016. "Autonomous Vehicles: Self-Driving Vehicles Enacted Legislation." December 12. http://www.ncsl.org/research/transportation/autonomous-vehicles-legislation.aspx#Enacted.

National Highway Traffic Safety Administration. 2016a. "2015 Motor Vehicle Crashes: Overview." August. https://crashstats.nhtsa.dot.gov/Api/Public/ViewPublication/812318.

National Highway Traffic Safety Administration. 2016b. "Federal Automated Vehicles Policy." September. https://www.transportation.gov/AV/federal-automated-vehicles-policy-september-2016.

National Highway Traffic Safety Administration. 2017a. "Automated Vehicles." February 15. https://www.nhtsa.gov/technology-innovation/automated-vehicles.

National Highway Traffic Safety Administration. 2017b. "Automated Driving Systems 2.0: A Vision for Safety." September. https://www.transportation.gov/av/2.0.

National Highway Traffic Safety Administration. 2018. "Preparing for the Future of Transportation: Automated Vehicles 3.0." October. https://www.transportation.gov/av/3.

National Highway Traffic Safety Administration. 2020. "Ensuring American Leadership in Automated Vehicle Technologies: Automated Vehicles 4.0." January. https://www.transportation.gov/av/4.

Obama, Barack. 2016. "Self-Driving, Yes, But Also Safe, Pittsburgh." *Post-Gazette*, September 19. http://www.post-gazette.com/opinion/Op-Ed/2016/09/19/Barack-Obama-Self-driving-yes-but-also-safe/stories.

Pearl, Tracy Hresko. 2017. "Fast & Furious: The Misregulation of Driverless Cars." *NYU Annual Survey of American Law* 73: 15–23.

Pidot, Justin R. 2015. "Governance and Uncertainty." *Cardozo Law Review* 37: 113–84.

Price, Derek J. de Solla. 1986. *Little Science, Big Science . . . and Beyond*. New York: Columbia University Press.

Rojas Rueda, David, et al. 2020. "Autonomous Vehicles and Public Health." *Annual Review of Public Health* 41: 329–45. https://papers.ssrn.com/sol3/papers.cfm?abstract_id=3570348.

Ruhl, J. B. 2005. "Regulation by Adaptive Management—Is It Possible?" *Minnesota Journal of Law, Science, and Technology* 7: 21–57.

Shanker, Ravi, et al. 2013. *Autonomous Cars: Self-Driving the New Auto Industry Paradigm*. New York: Morgan Stanley. https://orfe.princeton.edu/~alaink/smartdrivingcars/pdfs/nov2013morgan-stanley-blue-paper-autonomous-cars%ef%bc%9a-self-driving-the-new-auto-industry-paradigm.pdf.

Stigler, George J. 1971. "The Theory of Economic Regulation." *The Bell Journal of Economics and Management Science* 2: 3–21.

Sunstein, Cass R. 1995. "Problems with Rules." *California Law Review* 83: 953–1026.

Sunstein, Cass R. 2002. "Probability Neglect: Emotions, Worst Cases, and Law." *Yale Law Journal* 112: 61–107.

Sunstein, Cass R. 2005. *Laws of Fear: Beyond the Precautionary Principle*. New York: Cambridge University Press.

Sylvester, Douglas J., et al. 2009. "Not Again! Public Perception, Regulation, and Nanotechnology." *Regulation & Governance* 3: 165–85.

Thierer, Adam. 2016. *Permissionless Innovation: The Continuing Case for Comprehensive Technological Freedom*. Arlington: Mercatus Center. http://permissionlessinnovation.org/wp-content/uploads/2016/03/Thierer_Permissionless_web.pdf.

Turner, Fred. 2006. *From Counterculture to Cyberculture: Stewart Brand, the Whole Earth Network, and the Rise of Digital Utopianism*. Chicago: University of Chicago Press.

Wansley, Matthew T. 2016. "Regulation of Emerging Risks." *Vanderbilt Law Review* 69: 401–78.

Webb Yackee, Jason, and Susan Webb Yackee. 2012. "Testing the Ossification Thesis: An Empirical Examination of Federal Regulatory Volume and Speed, 1950–1990." *George Washington Law Review* 80: 1456–58.

Whitehead, Charles K. 2012. "The Goldilocks Approach: Financial Risk and Staged Regulation." *Cornell Law Review* 97: 1267–308.

Wood, Stephen G., Stephen Q. Wood, and Rachel A. Wood. 2006. "Whither the Precautionary Principle? An American Assessment from an Administrative Law Perspective." *American Journal of Comparative Law* 54: 581–613.

World Commission on the Ethics of Scientific Knowledge and Technology. 2005. *The Precautionary Principle*. http://unesdoc.unesco.org/images/0013/001395/139578e.pdf.

Chapter 9

Planes, Trains, and Flying Taxis

Ethics and the Lure of Autonomous Vehicles

Joseph Herkert, Jason Borenstein, and Keith W. Miller

BACKGROUND

With the creation and deployment of an increasing number of autonomous vehicles (AVs), the literature on the ethics of the technology continues to grow. In that literature, the term "autonomous vehicle" is typically used to refer to cars and trucks, but there is a much broader array of technologies under that label that warrants ethical analysis. In our previous work, we primarily focused on cars, which fits with the standard use of the term "autonomous vehicle." More specifically, we drew attention to the ethical responsibilities of engineers who are designing self-driving cars (Borenstein et al. 2017). We put forward the argument that engineers and other professionals need to consider ethical obligations that go beyond mere harm avoidance and legal compliance. In a subsequent publication, we sought to identify ethical issues that emerge when examining self-driving cars from a system-level perspective (Borenstein et al. 2019). Our aim was to articulate the importance of expanding the discussion from ethical issues at the level of an individual self-driving car (the level focused on when the trolley problem is applied to these issues) to include more complex concerns stemming from cars designed by different manufacturers with different levels of autonomy being embedded in large-scale sociotechnical systems. The broader, system-level focus reveals issues that are essentially invisible if we only restrict ourselves to the decisions that individual cars (and their developers) make.

In this chapter, the focus will be on AVs in a broad sense of the term; we seek to identify ethical issues pertaining to a mix of vehicles that include more than cars. A key reason for examining other types of AVs, in addition to cars, relates to how these vehicles add complexity to transportation systems, including when vehicles interact with one another. The ethical implications of the design, deployment, and use of AVs such as autonomous trains and flying taxis warrant systematic study. We aim to discuss the current status of AVs and some of the ethical issues associated with each category of vehicle. In addition, we will describe ethical complexities emerging from a system of multiple types of AVs interacting with one another and provide some analysis of why the general idea of autonomous transportation systems is so alluring.

AUTONOMOUS VEHICLE PROLIFERATION

From 2017 to 2019 the Bloomberg Philanthropies/Aspen Institute Initiative on Cities and Autonomous Vehicles identified ninety-six cities worldwide that were conducting or planned to conduct AV pilot projects, and another forty cities conducting preliminary studies of AVs (Bloomberg 2019). Of these 136 projects, 107 were located in North America or Europe. AV use cases for the pilots include 58 percent transit, 17 percent private auto, and 8 to 9 percent each for freight, ride for hire (that is, ride share), and paratransit (for people with disabilities); in some of the cities, multiple uses are being contemplated. The World Economic Forum (2018) commissioned a study of the city of Boston that considered eight transportation modes in three categories including conventional mass transit (bus/subway and commuter rail), personal car (including conventional and autonomous), and mobility-on-demand (conventional taxi/ride hailing, autonomous taxi, autonomous shared taxi, and autonomous minibus). As these reports indicate, governments and other entities are pursuing many different types of AVs including the personal car, taxi, ride share, freight (urban), and various forms of autonomous transit including minibuses and paratransit.

Other autonomous options being piloted or implemented, some of which are less publicized, include commercial short-haul trucks (Boudway 2019), long-haul trucks (Tomlinson 2019), mining trucks (Burton 2019), freight trains (RailFreight.com 2019), cargo ships (Marr 2019), and flying taxis/delivery vehicles (Whitham 2019). In addition, the long-sought goal of autonomous commercial airplanes continues to be pursued (Silver 2019).

Self-driving cars are by far the most publicized type of AV, including those for personal use and for car share or ride share services. Self-driving accidents involving human fatalities, including some in Florida, California, and Arizona, slowed but did not stop AV companies from seeking to deploy the technology. The AV industry is bullish on the technology with recent estimates of achieving "full autonomy" ranging from 2020 to 2030 (Litman 2019; Walker 2019). Yet in most cases, the predictions refer to only level 3 (a human driver needed for oversight and backup) or level 4 (fully autonomous only under certain conditions) vehicles from the SAE (2019) levels of driving automation. (SAE International, aka SAE, was formerly known as the Society of Automotive Engineers.)

Waymo is testing a commercial autonomous ridesharing service (Hawkins 2018) and GM in collaboration with DoorDash is implementing a food delivery service with its Cruise AV (Lekach 2019). Other *autonomous commercial delivery* vehicles, including cars and light trucks, are also being planned or piloted. These include a collaboration between Walmart and Gatick AI (Korosec 2019), and a venture in China featuring level 4 delivery vehicles (FABU 2019). The COVID-19 pandemic has placed heightened emphasis on the development of autonomous delivery vehicles (Ohnsman 2020), especially smaller, slow-moving versions (Borenstein et al. 2020a).

Autonomous shuttles and buses are attracting attention worldwide despite a 2017 accident in Las Vegas on the first day of a pilot (Park 2017). For example, in China testing began in 2017 on "nearly full-sized" buses (Ho 2017). Deployment of autonomous minibuses began in 2018 (Korosec 2018), and self-driving buses are expected to be rolled out in Beijing by 2022 (Xinhua 2019). Cities in other countries such as Stockholm, Sweden, are also conducting pilots (Krishnan 2018). An autonomous shuttle service recently started in Brooklyn (Hawkins 2019) and a similar service may launch in the metro Atlanta area (Keenan 2019).

Autonomous commercial trucks are also garnering widespread interest. According to *Business Insider* (Premack 2019),

> There are plenty of autonomous trucks on the road. TuSimple has a fleet of more than 50 trucks making three to five revenue-generating routes per day in Arizona. Waymo resumed testing its self-driving trucks in Phoenix, after ending the tests two years ago. Embark's trucks drove more than 124,000 automated miles last year. And Tesla has been spotted testing some autonomous semis.

As the COVID-19 pandemic seems to have tempered, interest in ridesharing AVs, many startups and new partnerships have formed to focus on autonomous trucks, such as Waymo's alliance with Daimler (Hawkins 2020).

Autonomous trains are drawing the attention of AV advocates (Trentesaux et al. 2018). For many years, urban passenger trains, some of which are fully automated, have operated in airports and throughout metropolitan areas. Since

2017, the world's first autonomous freight train has operated in Australia (Thompson 2017). Plans are underway for prototypes of fully autonomous passenger and freight trains in France in the early 2020s (Nussbaum and Mawad 2018).

Autonomous cargo ships are being planned throughout the world, with Norway leading the way (CBInsights 2018). Norway built the Yara Birkeland, an autonomous cargo ship for inland shipping (Paris 2017), but the COVID-19 pandemic might impact the ship's future (Haun 2020). Britain's Rolls-Royce also developed an autonomous naval ship (Ong 2017).

Flying taxis and other autonomous aircraft have attracted significant industry attention (Hornyack 2020). Rice and Winter (2019) state that

> several companies are developing fully autonomous aircraft, including Amazon and UPS, which want to use them for deliveries. Boeing and Airbus are designing self-flying air taxis, which would be used for flights of about 30 minutes and carry between two and four passengers, and have tested prototypes. A company called Volocopter has been testing autonomous air taxis in Germany since 2016 and plans to conduct test flights in downtown Singapore this year. Ridesharing giant Uber, helicopter maker Bell and many other companies are also expressing interest in similar vehicles.

Flying cars, which could be used as taxis or for other purposes, have been tested in Japan; they reportedly can hover over the ground for a short duration (Singh 2019; Boyd 2020). Korean automaker Hyundai, partnering with Uber, hopes to have models of flying cars on the market by 2028 (Park 2020). Sabrewing Aircraft is developing a vertical takeoff cargo drone (De Reyes 2020). On a smaller scale, autonomous delivery drones are nearing commercialization (de León 2020).

Autonomous airplanes have been under development for quite some time. For example, Reliable Robotics has collaborated with FedEx on developing full automation for small cargo planes (Hull 2020). In a sign of things to come, the commercial airline industry is likely to design future planes so that only a single human pilot is required in the cockpit (Falk 2017). Even though an autopilot system may eventually not require any direct human supervision, what may prevent the removal of human pilots from the commercial airlines for the foreseeable future is objections by passengers (and by the pilots).

Ethical Concerns

Many ethical issues emerge in relation to AVs. Some of them might be distinctive to a particular type of AV. For example, the relatively long stopping distance needed for a train may pose safety challenges, and flying taxis will pose challenges for air traffic control that are not as relevant for ground-based passenger cars. Yet in general, the issues we are discussing are common across multiple types of vehicles. A key ethical concern that has drawn much attention is privacy, including the potential sharing of passenger data; that issue has been extensively discussed by others (for example, Lin 2016 and Glancy 2012). For the purposes of this chapter, we will focus on several other key areas of concern mentioned shortly relating to AVs.

Large-Scale Job Loss and Dislocation

At least three key factors will influence the magnitude of the potential employment disruption due to AVs. The first is whether a human operator or passenger will be needed in the vehicle, or whether the vehicle will be fully autonomous without much, if any, human supervision. Given how many people the transportation sector employs, significant job loss could emerge depending on the design decisions that are made. The second key factor is which types of new job opportunities might emerge due to the existence of the various types of AVs. For instance, demand for programmers, software engineers, and hardware engineers might increase. The third factor is the extent to which employee retraining will be available for those who lose their jobs as a result of vehicle automation.

Underlying these ethically significant issues is the fact that economic considerations are a key motivating force behind the development of AVs, such as buses and trucks, because one of the chief operating expenses for these vehicles is labor. According to Abe (2019), "drivers' salaries and overhead labor costs are 56.7% of the total operating costs" for buses and rail in major metropolitan regions in Japan. Yet the widespread adoption of the technology could have huge, and potentially detrimental, impacts on the employment market. In the United States, over four million people are employed as bus drivers, delivery drivers, heavy truck drivers, taxi drivers, or chauffeurs (Center for Global Policy Solutions 2017). The Bureau of Labor Statistics in the US Department of Labor (2019b) estimates that there were approximately 687,200 bus drivers in the United States in 2016. Clearly, the potential for shifts in the labor market is immense.

AVs will also have a significant impact on employment in the airplane and trucking industries. In the United States alone, there were nearly 125,000 commercial pilots in 2016 (Bureau of Labor Statistics 2019a). Over 3.5 million people work as truck drivers in the United States, including individuals who drive commercial trucks for their full-time job (Day and Hait 2019). It is important to note that a unique aspect of retraining in this sector is that the median age of truck drivers is higher than it is for the average worker (Day and Hait 2019). The rail industry also employs a considerable number of people (AAR n.d.; Data USA n.d.), and those employees could experience massive job loss. Though AVs will be slower to catch on in other sectors such as shipping, employment in those sectors could also eventually be at risk. For example, according to the USDOT (2019), there were over 41,000 large cargo ships worldwide in 2016.

Traffic Safety

Much has been written about the safety of autonomous passenger cars, including our own work (Borenstein et al. 2017, 2019). Here we focus on a few areas where traffic safety becomes much more difficult to maintain due to the presence of other types of AVs. For example, autonomous long-haul trucks will likely travel in fleets (or "platoons"); this could be problematic because one vehicle (for example, due to tire failure) has the potential to cause large-scale accidents. Advocates might respond by claiming that vehicle-to-vehicle communication could help mitigate this problem by relaying traffic information between vehicles and by maintaining a safe driving distance. However, one of the alleged virtues of having a driverless fleet is the ability for the vehicles to travel closer together (Greenemeier 2017); this tension between safety and efficiency could result in the lack of sufficient time for breaking and the close proximity between the trucks could make an accident difficult to avoid.

In principle, autonomous trucks may work effectively in rural areas due to less population density, among other factors. But urban travel raises many technical and ethical challenges. For example, human drivers of other vehicles (for example, cars or bicycles) may be distracted by or even afraid of driverless trucks.

A significant problem for autonomous trains is their long stopping distances (Lavrinc 2013) and the short line of sight in many locations. Trains that travel cross-country also have a broader range of challenges (for example, falling trees, varied weather conditions, etc.) than urban or airport trains that may operate in a relatively static, and often controlled, environment. Clearly, the challenges for a train running underground, indoors at an airport are far simpler than a freight train moving across urban intersections protected by signs and gates. In less controlled environments, many complications could emerge such as adverse weather or an inability to detect or communicate with other AVs if, for example, those vehicles are crossing over train tracks.

Flying taxis, drones, and airplanes, especially if they operate in a fully autonomous mode, would pose a number of daunting safety challenges. Flying taxis and drones would pose dangers to larger aircraft. And as with other aircraft, collisions with birds are a nontrivial concern. According to the FAA (2018), over 14,400 "wildlife strikes" occurred at US airports in

2017. Also, during violent storms, mechanical failures, or other emergencies, passengers may become panicked and attempt to override a taxi's or plane's autonomous systems.

At least in the short term, human factors with partial automation could cause significant safety issues, as has already been seen with train operator distraction (AP 2010), "driver" distraction in autonomous cars (Borenstein et al. 2020b), and problems with pilot distraction/confusion when interacting with an autopilot (Carr 2014). The two crashes of Boeing 737 MAX aircraft illustrate how the failure of proper integration between an autonomous system and human operators can result in tragedy (Herkert et al. 2020; Bogaisky 2019).

Security of Commercial Deliveries and Freight

Autonomous commercial trucks and freight trains pose security risks to the transportation of freight and other items. Physical and cyberattacks are likely, especially in sparsely populated areas. Autonomous cargo ships may be particularly vulnerable to piracy if malicious actors do not have the fear of possible physical harm from human guards on board. Mechanical failure at sea could leave vessels stranded, and navigation errors could lead to politically tense situations, for example, when a ship accidentally sails into another country's territorial waters. The potential lack of human operators on board could decrease the options for managing or deescalating a dangerous situation on the waters. The isolation of ships on ocean voyages increases the risks to people and cargo if an autonomous ship goes awry.

Vehicle Vulnerabilities and Passenger Security

Passenger security is a significant ethical concern for autonomous ride-share vehicles, such as shuttles, buses, trains, flying taxis (Hornyack 2020), and airplanes, all of which could be subject to physical and cyberattacks (including "joy hacking" and terrorism). An AV could be the target of various kinds of "active" attacks such as a hostile actor trying to take control over a vehicle's autopilot. The vehicle could also be subject to "passive" attacks such as if lane markings (Ackerman 2019), street signs, or other external landmarks or markers are deliberately modified to confuse sensors. Similar to other types of AVs, the potential lack of human personnel on board could entail fewer opportunities to manage problems. For instance, something that might be easily corrected by a human (noticing that a stop sign has been covered in spray paint) could pose serious challenges for an autonomous system.

Passenger Well-Being, Access, and Social Justice Concerns

For the modes of autonomous transportation that are designed to carry passengers, their health and safety could be at risk due to the lack of a human monitor. For example, in the event a passenger becomes ill in a flying taxi, the lack of a human operator could be problematic. Or if a rider of an autonomous bus is unruly or displays threatening behavior to other passengers, there could be a serious worry when there is no driver to act as an authorized protector for the passengers.

A subset of the AVs discussed earlier could be integrated into public transit systems. A diverse range of passengers use public transit, including those with disabilities who may need physical assistance to get in and out of the vehicle. If that assistance is lacking, that raises concerns about access and social justice. Furthermore, autonomous passenger vehicles will need to be designed with features that enable a diverse range of users (including older adults and those with disabilities) to interface with the technology (Desmond 2020). For example, if a blind person is telling a flying taxi where to stop, the interface has to enable the user to interact reliably with the vehicle, especially if there is no human operator. More broadly, a significant investment of resources will be needed to ensure that the individuals inside and outside of AVs can safely and effectively interact with the technology.

Multiple Automated Transportation Systems

While the issues noted earlier are of concern when considering individual modes of transport (for example, autonomous trucks), they will be greatly magnified in a system that simultaneously includes multiple modes of automated transportation. Employment effects could be compounded due to limited opportunities to shift employees to similar jobs if various sectors are automated (for example, bus drivers will not shift to shuttle drivers if both jobs are disappearing). Providing safety and security for passengers, freight, and other goods will become much more complex due to the possibility of multiple system failures, including vulnerability to physical attacks (for example, terrorism, electromagnetic pulses from nuclear weapons, signage and signal vandalism), vulnerability to cyberattacks (for example, foreign or domestic hackers), and, more commonly, vulnerability to major system failures and common-mode failures (for example, power outages, internet outages, or natural disasters).

Systems for each vehicle type will need to process massive amounts of data. For instance, Zuboff (2019) claims that "a single autonomous car will generate 100 gigabytes per second." Vehicle-to-vehicle communication among different types of vehicles created by different manufacturers will be complicated, especially in the absence of local, national, and international standards. And standardization is something that AV companies are likely to resist for reasons such as they may have to redesign their products, share propriety information that they prefer remain secret, or because they may want to have a unique feature that separates them from competitors. Furthermore, other technical problems such as sensor interference among vehicle types could emerge. For example, as AVs become more common, the Light Detection and Ranging (LIDAR) sensors on different vehicles might interfere with one another (Eom et al. 2019).

The development of multiple autonomous transportation systems will require significant changes to infrastructure including the addition of facilities not currently needed. This would include, for example, fueling or charging stations for autonomous trucks, and space where flying taxis can take off and land. Regarding the latter, the need for landing areas generates numerous questions about land use priorities for city planners and others. The history of airport expansions includes many complaints related to social justice (Team Ecohustler 2019).

The use of various types of AVs also complicates decisions surrounding prioritization at intersections and crossings. Should autonomous trucks have priority over autonomous passenger vehicles? Should the decision be influenced by the size of truck, what it is carrying, and/or how many individuals are in a passenger vehicle? Also, how will nonoccupants (for example, pedestrians and cyclists) be protected and who has the primary responsibility to address this issue (Borenstein et al. 2020b)?

Macroethical concerns (Herkert 2005) will abound in a world of autonomous transportation systems. Due to the current and pervasive public skepticism about AVs (Liernert and Caspani 2019), the impacts on travel and tourism are uncertain. Public transportation systems may be downplayed in lieu of greater focus on individual transport; emphasizing the value of AVs may come at the cost of active transportation options (for example, walking or cycling) even for short trips, and thus impact public health.

An abundance of social justice concerns also emerges in relation to AVs (for example, due to priority of vehicle dispatching) (Mladenovic and McPherson 2016), and human autonomy may be challenged as more transportation choices are made by machines. Furthermore, the need for centralized control of transportation systems may grow with the increasing complexity and types of AVs. However, with centralized control of transportation by corporations and/or governments, the potential exists that the technology could be used for ethically dubious purposes (for example, public surveillance or predatory rates to ride in a vehicle). If efficiently and safely organized large-scale AV systems require centralized control, then the power differential between passengers and transportation providers may lead to ethical problems (Smith 2020).

All of the complications described will require time, effort, and money to untangle. An interesting question arises: Who will pay for dealing with those issues? If multinational corporations reap major profits for their AVs, will it be their responsibility to pay for the infrastructure necessary to make the resulting transportation systems safe and accessible? Or (more likely) will governments at various levels be expected to pick up the tab? The allocation of scarce public resources will likely divert funds and attention from other worthy concerns. Ethical, legal, and technical issues will abound at political boundaries as corporate, governmental, and public interests collide.

THE LURE OF AUTONOMOUS VEHICLES

Taking a broader view of the phenomena of AVs, it is prudent to consider the entire enterprise and why the vehicles seem so alluring. AVs are no different than various other emerging technologies that are perceived as a Technological Fix (a term coined by Weinberg [1966]) for social problems, including those that might not be directly related to transportation issues (for example, poverty). Embracing the latest Technological Fix is an approach that runs the risk of avoiding tackling such problems sincerely and directly. Related to this is the "Technocratic View of Progress" described by Leo Marx (1987), which holds that improved technology is an end in itself that will be accompanied by general progress, rather than viewing technology as a means to achieving social ends.

It is likely that "technological somnambulism," a concept described by Langdon Winner (1983), has and will continue to plague human thinking in relation to autonomous technology. In part, technological somnambulism refers to a lack of critical thinking about how technology is going to reshape human life and the physical environment to keep up with technological advances. Winner (1983) states that "it is reasonable to suppose that a society thoroughly committed to making artificial realities would have given a great deal of thought to the nature of that commitment," but he notes that this does not often occur. In this case, it is not fully clear that each stakeholder involved in the development and deployment of AVs has made a serious enough commitment to analyzing the technology's societal impacts.

The pervasive "move fast and break things" mentality has been entrenched within many technology companies during the digital age (Vardi 2018). At a surface level, the message has appeal; to grow and be on the cutting edge of innovation, one may need to take bold risks. Yet what this mentality often hides or masks is the distribution of the harms tied to those risks. If an autonomous flying taxi crashes, the passengers might die, as might people on the ground. The developers and vendors of the flying taxi are unlikely to be victims, but they are certainly likely to have benefited from that flying taxi's sale and deployment. The advocates of the move fast and break things mentality often embrace it because they are unlikely to be the ones who are going to experience the lion's share of the resulting harms.

As mentioned previously, the widespread adoption of AVs is likely to have serious and disruptive effects on the workforce. Compounding this is that other types of robots, and artificial intelligence more generally, might replace entire sectors of human labor (for example, Shewan 2017). It is not clear if governments or other entities have grappled with the ripple effects this could have on communities. There are beginnings of discussions in the United States about how to address the issue of automation generally including with retraining programs (for example, McKinsey Global Institute 2017). Interestingly, Andrew Yang, a former candidate for the US presidency in 2020, made it his defining campaign issue to provide a universal basic income as an approach to mitigating the impacts of automation.

Workforce disruption is not a surprising consequence of AVs because a major rationale for AVs is to replace human labor. While this is usually cast in economic terms, another lure of AVs may well be the desire to eliminate "human error." A common response to technological accidents, both by engineers and others, is to blame the human user or operator rather than considering shortcomings in the design or manufacturing of the technology (Holden 2009). This takes on special significance for AVs where it may no longer be possible to blame a human user/operator. One response to this has been to shift more responsibility for their safety onto pedestrians, cyclists, and other nonusers of AVs. "The ethical significance of this shifting of responsibility is clear. Surely all of the people who share the road have responsibilities for their own safety and the safety of others. But it would be ethically problematic if the developers of the new technology suggest that pedestrians, not car manufacturers, are primarily responsible for pedestrian safety in situations when automated vehicles mingle with pedestrians and other non-occupants" (Borenstein et al. 2020b).

Close (2016) describes the desire for new and better gadgets as being like a "religious experience," which hints at the notion that some have an irrational enthusiasm for the latest technologies. Stivers and Stirk (2001) use the phrase "technology as magic" to illustrate how a technology's characteristics are viewed with great optimism, especially during the early stages of Gartner's hype cycles (Linden and Fenn 2003). One might see this kind of enthusiasm in Wiseman (2018), who claims, "In an era of autonomous vehicles, rails are obsolete."

When a company's financial interests are closely tied to mass acceptance of a technology (and this is clearly the case for AVs), it can spend vast amounts of advertising and lobbying dollars to emphasize benefits and downplay risks and costs. Yet relatively small investments in improving public transportation (for example, trains and buses) are likely to have more immediate and lasting improvements in safety and environmental conditions than relatively large investments in building an entirely new system infrastructure for an all automated vehicle future. Moreover, even modest improvements in public transportation could have significant positive effects for less well-off passengers; current plans for autonomous AVs tend to favor those who can afford to buy them. Mass transit improvements could include AVs (Bischoff et al. 2017), but enhanced public transportation is certainly not the vision that has captured people's attention.

Trains and buses (even when they are fully automated) do not have the wow factor of a personalized, autonomous automobile. Individuals do not personally own metropolitan light rail systems; an increase in their social status does not usually follow from riding the train to work. There are technologies other than an autonomous personal vehicle that are less expensive and potentially more effective at avoiding traffic gridlock or achieving other mobility-related goals. Yet at least for now, it appears that the more glamorous (and more profitable for corporations) vehicles are gaining the most attention. It will require much effort to counter the forces determined to enhance the wow factor of autonomous personal vehicles.

CONCLUSIONS AND RECOMMENDATIONS

AVs of all types, for land, sea, and air, are capturing the attention of innovators and investors and the imagination of portions of the public. While the motivations behind the lure of AVs may be more hidden and complex than the developers of the technology realize or acknowledge, the momentum toward implementing these technologies has without a doubt begun. With that in mind, we offer several recommendations aimed at guiding the design and development of AVs to increase the likelihood that it proceeds in an ethical and socially responsible manner.

1. All stakeholders need to acknowledge that ethical challenges beyond abstract trolley problems will emerge from individual autonomous transportation modes and are even more daunting for integrated autonomous transportation systems. There is a critical need to move the unit of analysis from the individual vehicle or type of vehicle to a system-level

perspective that appreciates the complexity of the sociotechnical systems within which AVs will function.
2. While it is an encouraging sign that nearly half of the urban AV projects focus on transit (Bloomberg 2019), most of the hype around AVs tends to focus on cars, trucks, and more esoteric technologies such as flying taxis. The funding and regulation of AVs should prioritize mass transit, including buses, shuttles, and trains. This will ensure the benefits of AVs are more widely shared across a community and not just for affluent segments of society or hobbyists.
3. Job loss and dislocation should be addressed in parallel with the development of AVs; educational and retraining opportunities must be in place before AVs become pervasive.
4. A primary responsibility of AV technology developers is to protect the safety and security of both AV occupants and nonoccupants. Regulations to this effect should be in place before AVs take to the road in large numbers.
5. Standards, informed by technical, ethical, and other relevant concerns, are needed to ensure that multiple vehicle types developed by multiple manufacturers are able to communicate and operate in a safe and efficient manner. The creation of such standards could have the added benefit of being a measure against which AV developers could be held accountable.
6. Substantial vehicle testing of AVs should take place in controlled environments before proceeding to public roads; this testing should include multiple vehicle types (for example, autonomous cars and autonomous buses) and variable operating conditions.

REFERENCES

Abe, Ryosuke. 2019. "Introducing Autonomous Buses and Taxis: Quantifying the Potential Benefits in Japanese Transportation Systems." *Transportation Research Part A: Policy and Practice* 126: 94–113.

Ackerman, E. 2019. "Three Small Stickers in Intersection Can Cause Tesla Autopilot to Swerve into Wrong Lane." *IEEE Spectrum*, April 1. https://spectrum.ieee.org/cars-that-think/transportation/self-driving/three-small-stickers-on-road-can-steer-tesla-autopilot-into-oncoming-lane.

Associated Press. 2010. "NTSB Blames Texting in Calif. Rail Crash." NBCNews.com, January 21. http://www.nbcnews.com/id/34978572/ns/us_news-life/t/ntsb-blames-texting-deadly-calif-rail-crash/#.WnMl66inFPZ.

Association of American Railroads (AAR). n.d. "Freight Rail in Your State." https://www.aar.org/data-center/railroads-states/.

Bischoff, J., I. Kaddoura, M. Maciejewski, and K. Nagel. 2017. "Re-defining the Role of Public Transport in a World of Shared Autonomous Vehicles." In Symposium of the European Association for Research in Transportation (hEART).

Bloomberg Philanthropies, The Aspen Institute. 2019. "Initiative on Cities and Autonomous Vehicles." https://avsincities.bloomberg.org/.

Bogaisky, Jeremy. 2019. "Boeing 737 MAX Raises Concerns Over How FAA Will Ensure the Safety of Autonomous Aircraft." *Forbes*, April 24. https://www.forbes.com/sites/jeremybogaisky/2019/04/24/boeing-737-max-faa-autonomous-aircraft.

Borenstein, Jason, Joseph Herkert, Yvette Pearson, and Keith Miller. 2020a. "Autonomous Vehicles and Pandemic Response: Ethical Challenges and Opportunities." Presented at the Online Forum on Philosophy, Engineering and Technology (fPET 2020).

Borenstein, J., J. Herkert, and K. Miller. 2020b. "Autonomous Vehicles and the Ethical Tension Between Occupant and Non-Occupant Safety." *Journal of Sociotechnical Critique* 1 (1): 1–14. https://doi.org/10.25779/5g55-hw09.

Borenstein, Jason, Joseph Herkert, and Keith Miller. 2019. "Self-Driving Cars and Engineering Ethics: The Need for a System Level Analysis." *Science and Engineering Ethics* 25 (2): 383–98.

Borenstein, Jason, Joseph Herkert, and Keith Miller. 2017. "Self-Driving Cars: Ethical Responsibilities of Design Engineers." *IEEE Technology and Society Magazine* 36 (2): 67–75.

Boudway, Ira. 2019. "Self-Driving Trucks Will Carry Mail in U.S. for the First Time." Bloomberg.com. https://www.bloomberg.com/news/articles/2019-05-21/self-driving-trucks-will-carry-mail-in-u-s-for-the-first-time.

Boyd, J. 2020. "Japan on Track to Introduce Flying Taxi Services in 2023." *IEEE Spectrum*, September 4. https://spectrum.ieee.org/cars-that-think/aerospace/aviation/japan-on-track-to-introduce-flying-taxi-services-in-2023.

Brey, P. A. 2012. "Anticipatory Ethics for Emerging Technologies." *NanoEthics* 6 (1): 1–13.

Burton, Melanie. 2019. "Rio Tinto to Buy Autonomous Mining Truck Fleet from Caterpillar." *Reuters*, May 6. https://www.reuters.com/article/us-australia-mining-autonomous/rio-tinto-to-buy-autonomous-mining-truck-fleet-from-caterpillar-idUSKCN1SD01S.

Carr, N. 2014. *The Glass Cage: Automation and Us*. New York: W. W. Norton & Company.

CBInsights 2018. "Massive Cargo Ships Are Going Autonomous. Here Are The Companies & Trends Driving The Global Maritime Industry Forward." https://www.cbinsights.com/research/autonomous-shipping-trends/.

Center for Global Policy Solutions. 2017. "Stick Shift: Autonomous Vehicles, Driving Jobs, and the Future of Work." http://globalpolicysolutions.org/report/stick-shift-autonomous-vehicles-driving-jobs-and-the-future-of-work/.

Close, Kerry. 2016. "Why Buying the Latest Apple Gadget Can Feel Like a Religious Experience." *Money*, May 6. http://money.com/psychology-why-want-gadgets-iphone.

Data USA. n.d. "Locomotive Engineers & Operators." https://datausa.io/profile/soc/locomotive-engineers-operators.

Davies, Alex. 2016. "Inside Uber's Plan to Take Over the Skies with Flying Cars." *Wired*, October 27. https://www.wired.com/2016/10/uber-flying-cars-elevate-plan/.

Day, Jennifer Cheeseman, and Andrew W. Hait. 2019. "Number of Truckers at All-Time High." US Census Bureau, June 6. https://www.census.gov/library/stories/2019/06/america-keeps-on-trucking.html.

de León, C. 2020. "Drone Delivery? Amazon Moves Closer with F.A.A. Approval." *New York Times*, August 31. https://www.nytimes.com/2020/08/31/business/amazon-drone-delivery.html.

De Reyes, E. 2020. "Can Cargo Drones Solve Air Freight's Logjams? A Drone Startup Says Its Big Vertical-Takeoff Flier Would Be Quick to Land, Load, and Take Off Again." *IEEE Spectrum* 57 (6): 30–35.

Desmond, N. 2020. "Hailing a Self-Driving Taxi When Blind. Learn How Waymo Answers That Challenge at Sight Tech Global." *TechCrunch*, September 21. https://techcrunch.com/2020/09/21/hailing-a-self-driving-taxi-when-blind-learn-how-waymo-answers-that-challenge-at-sight-tech-global/.

Dougherty, C. 2017. "Self-Driving Trucks May Be Closer Than They Appear." *New York Times*, November 13. https://www.nytimes.com/2017/11/13/business/self-driving-trucks.html.

Eom, Jeongsook, Gunzung Kim, and Yongwan Park. 2019. "Mutual Interference Potential and Impact of Scanning Lidar according to the Relevant Vehicle Applications." Proc. SPIE 11005, Laser Radar Technology and Applications XXIV, 110050I, https://doi.org/10.1117/12.2518643.

FABU Technology. 2019. "In China, First Self-Driving Trucks to Begin Commercial Deliveries Utilizing FABU Technology." https://www.prnewswire.com/news-releases/in-china-first-self-driving-trucks-to-begin-commercial-deliveries-utilizing-fabu-technology-300805458.html.

Falk, D. 2017. "Self-Flying Planes May Arrive Sooner Than You Think. Here's Why." *Mach*, October 11. https://www.nbcnews.com/mach/science/self-flying-planes-may-arrive-sooner-you-think-here-s-ncna809856.

Glancy, D. J. 2012. "Privacy in Autonomous Vehicles." *Santa Clara Law Review* 52: 1171.

Greenemeier, Larry. 2017. "Driverless Big Rigs Nearly Road Ready—That's a Big 10–4." *Scientific American*. https://www.scientificamerican.com/article/driverless-big-rigs-nearly-road-ready-mdash-thats-a-big-10-4/.

Haun, Eric. 2020. "Yara Birkeland Project Paused Due to COVID-19." *MarineLink*. https://www.marinelink.com/news/yara-birkeland-project-paused-due-covid-478386.

Hawkins, Andrew J. 2020. "Waymo and Daimler Are Teaming Up to Build Fully Driverless Semi Trucks." *The Verge*, October 27. https://www.theverge.com/2020/10/27/21536048/waymo-daimler-driverless-semi-trucks-cascadia-freightliner.

Hawkins, Andrew J. 2019. "New York City's First Self-Driving Shuttle Service Is Now Open for Business." *The Verge*, August 6. https://www.theverge.com/2019/8/6/20755163/new-york-city-self-driving-shuttle-service.

Hawkins, Andrew J. 2018. "Riding in Waymo One, the Google Spinoff's First Self-Driving Taxi Service." *The Verge*. https://www.theverge.com/2018/12/5/18126103/waymo-one-self-driving-taxi-service-ride-safety-alphabet-cost-app.

Herkert, J. R. 2005. "Ways of Thinking About and Teaching Ethical Problem Solving: Microethics and Macroethics in Engineering." *Science and Engineering Ethics* 11 (3): 373–85.

Herkert, Joseph, Jason Borenstein, and Keith Miller. 2020. "The Boeing 737 MAX: Lessons for Engineering Ethics." *Science and Engineering Ethics* 26 (6): 2957–74. https://doi.org/10.1007/s11948-020-00252-y.

Ho, V. 2017. "Self-Driving Buses Are Being Tested in China and They're The Largest of Their Kind Yet." *Mashable*, December 4. https://mashable.com/2017/12/04/self-driving-bus-china/#LfZ1wqejbOqo.

Holden, R. J. 2009. "People or Systems? To Blame Is Human. The Fix Is to Engineer." *Professional Safety* 54 (12): 34.

Hornyack, T. 2020. "The Flying Taxi Market May Be Ready for Takeoff, Changing the Travel Experience Forever." CNBC, March 7. https://www.cnbc.com/2020/03/06/the-flying-taxi-market-is-ready-to-change-worldwide-travel.html.

Hull, D. 2020. "SpaceX Veterans Pass Milestone in Self-Flying Plane Race." *Bloomberg*, August 26. https://www.bloomberg.com/news/articles/2020-08-26/spacex-veterans-pass-milestone-in-self-flying-plane-race.

Keenan, Sean. 2019. "Pioneering Autonomous Shuttle to Begin Service at Doraville Development This Spring." *Curbed*, March 19. https://atlanta.curbed.com/2019/3/19/18272103/autonomous-shuttle-doraville-transit-oriented-marta-last-mile-connectivity.

Korosec, Kirsten. 2018. "Baidu Just Made Its 100th Autonomous Bus Ahead of Commercial Launch in China." *TechCrunch*, July 3. https://techcrunch.com/2018/07/03/baidu-just-made-its-100th-autonomous-bus-ahead-of-commercial-launch-in-china/.

Korosec, Kirsten. 2019. "Self-Driving Delivery Van Startup Gatki AI Comes Out of Stealth with Walmart Partnership." *TechCrunch*. https://techcrunch.com/2019/06/06/self-driving-delivery-van-startup-gatik-ai-comes-out-of-stealth-with-walmart-partnership/.

Krishnan, R. 2018. "Self-Driving Shuttle Buses Hit the Streets of Stockholm." *New Atlas*, January 25. https://newatlas.com/ericsson-self-driving-buses/53126/.

Lavrinc, Damon. 2013. "It's Not a Lack of Technology That's Keeping Trains from Going Driverless." *Wired*, April 26. https://www.wired.com/2013/04/why-arent-trains-autonomous/.

Lekach, Sasha. 2019. "Self-Driving Robots and Cars Deliver Food to Techies and College Students." *Mashable*. https://mashable.com/article/cruise-gm-doordash-autonomous-food-delivery/.

Liernert, Paul, and Maria Caspani. 2019. "Americans Still Don't Trust Self-Driving Cars, Reuters/Ipsos Poll Finds." *Reuters*, April 1. https://www.reuters.com/article/us-autos-selfdriving-poll/americans-still-dont-trust-self-driving-cars-reuters-ipsos-poll-finds-idUSKCN1RD2QS.

Lin, P. 2016. "Why Ethics Matters for Autonomous Cars." In *Autonomous Driving*, 69–85. Berlin: Springer.

Linden, Alexander, and Jackie Fenn. 2003. *Understanding Gartner's Hype Cycles*. Strategic Analysis Report No. R-20-1971. Gartner, Inc.

Litman, Todd. 2019. "Autonomous Vehicle Implementation Predictions: Implications for Transport Planning." Victoria Transport Policy Institute. https://www.vtpi.org/avip.pdf.

Marr, Bernard. 2019. "The Incredible Autonomous Ships of the Future: Run by Artificial Intelligence Rather Than a Crew." *Forbes*. https://www.forbes.com/sites/bernardmarr/2019/06/05/the-incredible-autonomous-ships-of-the-future-run-by-artificial-intelligence-rather-than-a-crew/#7bdd88936fbf.

Marx, L. 1987. "Does Improved Technology Mean Progress." *Technology Review* 90 (1): 33–41.

McKinsey Global Institute. 2017. "A Future That Works: Automation, Employment, and Productivity." https://www.mckinsey.com/featured-insights/digital-disruption/harnessing-automation-for-a-future-that-works/de-DE.

Mladenovic, M. N., and T. McPherson. 2016. "Engineering Social Justice into Traffic Control for Self-Driving Vehicles?" *Science and Engineering Ethics* 22 (4): 1131–49.

Nussbaum, Ania, and Marie Mawad. 2018. "France Mulls Autonomous Trains." *Transport Topics*, September 12. https://www.ttnews.com/articles/france-mulls-autonomous-trains.

Ohnsman, A. 2020. "Coronavirus Scrambles Self-Driving Race, Pushing Nuro's Delivery Bots to Front of the Pack." *Forbes*, July 9. https://www.forbes.com/sites/alanohnsman/2020/07/09/coronavirus-scrambles-self-driving-race-pushing-nuros-delivery-bots-to-front-of-the-pack/#6875998e4a15.

Ong, T. 2017. "Rolls-Royce Has Plans for an Autonomous Naval Ship." *The Verge*, September 13. https://www.theverge.com/2017/9/13/16300866/rolls-royce-autonomous-ship-navy.

Paris, C. 2017. "Norway Takes Lead in Race to Build Autonomous Cargo Ships." *The Wall Street Journal*, July 22. https://www.wsj.com/articles/norway-takes-lead-in-race-to-build-autonomous-cargo-ships-1500721202.

Park, K. 2020. "Hyundai Confident on Flying Cars, Steps Up Plans for Full Lineup." *Bloomberg*, October 5. https://www.bloomberg.com/news/articles/2020-10-05/hyundai-confident-on-flying-cars-steps-up-plans-for-full-lineup.

Park, M. 2017. "Self-Driving Bus Involved in Accident on Its First Day." Cnn.com, November 9. http://money.cnn.com/2017/11/09/technology/self-driving-bus-accident-las-vegas/index.html.

Premack, Rachel. 2019. "A Little-Known Trucking Startup Just Beat Tesla and Waymo to Run Driverless Semi-Trucks on the Open Road." *Business Insider*, June 26. https://www.businessinsider.com/starsky-beat-tesla-waymo-driverless-unmanned-trucks-2019-6.

Railfreight.com. 2019. "Rollout World's First Driverless Freight Train Network Complete." https://www.railfreight.com/technology/2019/01/04/rollout-worlds-first-driverless-freight-train-network-complete/?gdpr=accept.

Rice, Stephen, and Scott Winter. 2019. "Despite Passenger Fears, Automation Is the Future of Aviation." *Discover*, March 27. http://blogs.discovermagazine.com/crux/2019/03/27/plane-aviation-automation-autopilot/.

SAE International. 2019. "Levels of Driving Automation." https://www.sae.org/news/2019/01/sae-updates-j3016-automated-driving-graphic.

Shewan, Dan. 2017. "Robots Will Destroy Our Jobs—And We're Not Ready for It." *The Guardian*, January 11. https://www.theguardian.com/technology/2017/jan/11/robots-jobs-employees-artificial-intelligence.

Silver, David. 2019. "The Tesla Autopilot Model of Autonomous Flight." *Forbes*. https://www.forbes.com/sites/davidsilver/2019/05/06/the-tesla-autopilot-model-of-autonomous-flight/#64f2217c440c.

Singh, Karishma. 2019. "Japanese Flying Car Hovers for a Minute during Test Flight." *Reuters*, August 6. https://www.reuters.com/article/us-japan-flying-car/japanese-flying-car-hovers-for-a-minute-during-test-flight-idUSKCN1UW0UR.

Smith, Bryant Walker. 2020. "Ethics of Artificial Intelligence in Transport." In *The Oxford Handbook of Ethics of Artificial Intelligence*, edited by Markus Dubber, Frank Pasquale, and Sunit Das. Oxford University Press.

Stivers, Richard, and Peter Stirk. 2001. *Technology as Magic: The Triumph of the Irrational*. A&C Black.

Team Ecohustler. 2019. "The People Versus Airport Expansion." https://ecohustler.com/technology/the-people-versus-airport-expansion/.

Thompson, A. 2017. "Self-Driving Freight Trains Are Now Traveling the Rails Without a Human on Board." *Popular Mechanics*, October 2. https://www.popularmechanics.com/technology/infrastructure/news/a28470/new-driverless-train-could-reinvent-shippin/.

Tomlinson, Chris. 2019. "Autonomous Semi-Trucks Coming to I-10 Soon." *Houston Chronicle*. https://www.houstonchronicle.com/business/columnists/tomlinson/article/Autonomous-semi-trucks-coming-to-I-10-soon-13707258.php.

Trentesaux, D., R. Dahyot, A. Ouedraogo, D. Arenas, S. Lefebvre, W. Schön, B. Lussier, and H. Cheritel. 2018. "The Autonomous Train." In 2018 13th Annual Conference on System of Systems Engineering (SoSE), June (514–20). IEEE.

US Department of Labor, Bureau of Labor Statistics. 2019a. *Occupational Outlook Handbook, Airline and Commercial Pilots*. https://www.bls.gov/ooh/transportation-and-material-moving/airline-and-commercial-pilots.htm.

US Department of Labor, Bureau of Labor Statistics. 2019b. *Occupational Outlook Handbook, Bus Drivers*.

US Department of Transportation, Bureau of Transportation Statistics. 2019. *Number and Size of the U.S. Flag Merchant Fleet and Its Share of the World Fleet*. https://www.bts.gov/content/number-and-size-us-flag-merchant-fleet-and-its-share-world-fleet.

US Department of Transportation, Federal Aviation Administration (FAA). 2018. "Frequently Asked Questions and Answers." https://www.faa.gov/airports/airport_safety/wildlife/faq.

Vardi, Moshe Y. 2018. "Move Fast and Break Things." *Communications of the ACM* 61 (9): 7.

Walker, Jon. 2019. "The Self-Driving Car Timeline—Predictions from the Top 11 Global Automakers." https://emerj.com/ai-adoption-timelines/self-driving-car-timeline-themselves-top-11-automakers.

Weinberg, A. M. 1966. "Can Technology Replace Social Engineering?" *Bulletin of the Atomic Scientists* 22 (10): 4–8.

Whitham, Ryan. 2019. "Boeing Tests Autonomous Flying Taxi." https://www.extremetech.com/extreme/284461-boeing-tests-autonomous-flying-taxi.

Winner, Langdon. 1983. "Technologies as Forms of Life." In *Epistemology, Methodology and the Social Sciences*, edited by Robert S. Cohen and Marx W. Wartofsky. Kluwer Academic Publishers.

Wiseman, Yair. 2018. "In an Era of Autonomous Vehicles, Rails Are Obsolete." *International Journal of Control and Automation* 11 (2): 151–60.

World Economic Forum. 2018. "Reshaping Urban Mobility with Autonomous Vehicles: Lessons from the City of Boston." http://www3.weforum.org/docs/WEF_Reshaping_Urban_Mobility_with_Autonomous_Vehicles_2018.pdf.

Xinhua. 2019. "Beijing Plans to Launch Self-Driving Buses in 2022." ChinaDaily.com.cn. http://www.chinadaily.com.cn/a/201901/11/WS5c383944a3106c65c34e3f0f.html.

Zuboff, Shoshana. 2019. *The Age of Surveillance Capitalism: The Fight for a Human Future at the New Frontier of Power*. New York: PublicAffairs.

Chapter 10

Experiencing the Future

A Phenomenological Exploration of Automated Vehicles

Ike Kamphof and Tsjalling Swierstra

As in crime, there are "usual suspects" in technology ethics. If a new technology appears on the horizon and attracts sufficient attention to be deemed worthy of ethical assessment, it gets routinely scanned for risks. Risks, in practice, mean hazards to our health, safety, and environment. Increasingly common is also an assessment in terms of accountability, distributive justice, and inclusivity, while digital products are routinely scanned for data security and privacy issues. Automated vehicles are no exception. By far the most ethical literature on this emerging technology concentrates on these "usual suspects" (cf. Thompson n.d.; Santoni De Sio 2016). Think for instance of all the ink spent on—or, more accurately, the clicks devoted to—trolley dilemmas (Awad et al. 2018).

What holds true for the ethical literature on self-driving vehicles applies even more to ethical policies. The most fully developed one is probably the ethical code recently adopted by Germany (Luetge 2017). In 2016, the German Federal Minister of Transport and Digital Infrastructure appointed a national ethics committee for automated and connected driving. In June 2017, this committee presented a code of ethics (BMVI 2017a), which was accepted by the German government as foundation for future legislation and legal standards (BMVI 2007b). Seventeen of the twenty rules constituting this code deal with safety, harm, and (conditions for, and attribution of) accountability. The remaining three deal with surveillance, cybersecurity, and privacy. Distributive justice is not part of the code—probably because that is considered politically sensitive.

The German ethical code illustrates how the focus in the ethics of technology and in policy making is most often on so-called hard impacts. Hard impacts are typically instances of (bodily) harm; the impacts are clearly observable and often quantifiable, and they are causally connected to the technology (Swierstra 2015). In the context of automated vehicles, the popular trolley problem is a perfect example, as it scores high on all three counts: everyone agrees that casualties are bad, the casualties can be counted, and no one will argue that the casualties are not caused by the trolley.

There are sound reasons for this focus. First of all, John Stuart Mill's emphasis on avoiding harm has provided pluralist societies with a narrow, but very solid, moral foundation (Mill 1859, 21–22). The language of objective and quantifiable fact, in many cases, provides a similarly solid basis for public deliberation, policymaking, and regulation. The demonstrability of a direct causal link between technology and impact facilitates the organization of an accountability system that incentivizes a variety of heterogeneous actors to behave at their best.

However, it is also becoming increasingly clear that this focus on the hard impacts of technology may be too narrow to deal adequately with new and emerging technologies (Swierstra 2015). The assumption seems to be that as long as a technology doesn't explode, poison, or pollute—all is well. That, of course, is not true. Even seemingly harmless technology can instigate wide-ranging changes in our lives and societies that are not value neutral, and so may be considered good or bad. A new technology, like self-driving cars, has the potential to reshape *lived experience* in many ways. It may affect how we relate to the outside world of things and events, to the social world of relations, and to the subjective world of inner experience.

Let us illustrate such "soft" impacts of technology with a well-known example. If in the case of smartphones one would focus exclusively on harm, justice, or privacy, one would completely miss how this mundane device has thoroughly altered the way people experience time (for example, acceleration) and space (for example, virtual networks), how they act (for example, conducting private conversations in public), how they interact (for example, FOMO), and how they relate to themselves (for example, "quantified self," taking "selfies").

In this chapter we want to draw attention to such "soft" impacts of self-driving cars. Soft impacts differ from hard impacts insofar as they are *qualitative* rather than quantitative; they don't revolve around uncontroversial instances of harm but are instead ethically *unclear or ambiguous*. Rather than being a direct, causal consequence of the new technology, they are *coproduced* by both technical and social factors, including practices developed by users. Together these three features can make it difficult and contentious to attribute responsibility when something goes wrong. As a result, agents in industry or government find them hard to handle and often choose to ignore them—as is illustrated by the German Code mentioned earlier. We will argue that "hard to handle" does not mean that soft impacts don't need our attention and care. While most people, luckily, will be spared the hard impacts of self-driving vehicles, none of them will escape their soft impacts—comparable to how today's users and nonusers all have to come to terms with the soft impacts of smartphones.

How will the presence of self-driving vehicles change the ways people make their choices and act in the world? How will it affect their lived experience? Needless to say, this is impossible to predict. But it is possible to prepare for the future by employing our informed imagination. When the future takes shape, we may then be in a better position to deal with its challenges and affordances. To prepare ourselves for an unpredictable future, we should train our diagnostic skills. Only then will we be able to detect, articulate, discuss, and coshape the manifold ways technologies influence individual and shared lives on the *level of lived, mundane, experience.*

Methodologically we draw on phenomenology, the branch of philosophy that pays systematic attention to the field of everyday, lived experience. In this contribution, we will present and discuss four phenomenological anecdotes that suggest possible experiences in a world where self-driving cars have become the norm. These anecdotes are not predictions but aim to sensitize our discussions of a future with automated vehicles to effects other than the hard impacts focused on in the current literature. Before we present this work, we briefly discuss how we understand phenomenology as a methodology to explore not current but future practice.

A PHENOMENOLOGY OF (FUTURE) LIVED EXPERIENCE

Phenomenology studies how the world is given to us in prereflective experience, before we name and conceptualize it. It articulates this "lived experience" as part of practical contexts in which we perceive and are engaged with things in particular ways (van Manen 2014, 13). For instance, while for a mechanic a car, whether automated or conventional, may appear as transparent, to be fixed by the exercise of personal skill, the same car may be enigmatic in its working but serve as a symbol of success for a young professional, or as a cherished means of independence for an elderly user. In each case, the meaning of the car and identity of the user emerge together within a specific practical engagement. In short, phenomenology sees subjects and objects, identities and meanings, always in relation to each other.

In our case, a phenomenological approach can help to bring to light how a world where automated vehicles abound is lived by direct and indirect users. To what practices do these vehicles give rise and how do these impact the relationships of cars with their users and also nonusers? How does the use of automated vehicles shape how (non)users experience themselves and their surroundings?

Before we can start, however, we have to clarify one methodological issue. Being directed at the "living now" (van Manen 2014, 34), a phenomenological approach is normally understood as "radically empirical" (Ihde 2012, 16) and opposed to "free-floating constructions" (15). This raises problems in the case of the phenomenon we are interested in: how fully automated vehicles may give rise to new practical, meaning-giving contexts. As this largely lies in the future, a degree of imaginative speculation is unavoidable. In developing a "phenomenology of the future," we nevertheless remain true to the core principles of a phenomenology of practice, as developed by van Manen (2014), and the experimental postphenomenology of technological mediation elaborated by Ihde (2012). These are: a direction at the phenomenon in the fullness of its possibilities and a focus on the main forms of human experience in which these possibilities come to light. To secure that our speculative approach remains grounded in empirical phenomenology, we make use of some established rules for the presentation of phenomenological data. Let us expand on this.

The phenomenon we focus on are fully automated vehicles and the practical, meaning-giving contexts they give rise to. This phenomenon has not yet fully materialized, but neither is it a mere idiosyncratic fantasy. It is envisioned by engineers, promised by enthusiasts, feared by critics, desired or refused by citizens, debated by ethicists, etc. All of these anticipations contain, explicitly or implicitly, a variety of potential appearances of the phenomenon and of possible experiences of it, for example, as a source of safety or comfort or as a potential danger to health or privacy. This is the concrete phenomenon focused on in this chapter, not as sensually present here and now but in its "absences," or as imagined (Sokolowski 2000, 37).

When we normally speak about our experiences, our accounts tend to be shot through by interpretations of what the phenomenon is and explanations of why it is the way it is. Phenomenology calls this the "natural attitude." Phenomenological observation takes a step back from these—largely sedimented—meanings to observe the phenomenon anew, or as van Manen (2014) calls it, in a state of "wonder" (26). This stepping back—or "epoche"—discloses a phenomenon as a manifold and the experience of it as a richness of possibilities, of which only some come to explicit awareness in what we normally understand as "our experience." In placing ourselves, as researchers, in the position of "observing" the—admittedly largely imaginary—manifold appearance of automated vehicles, we aim to loosen the limited accounts and sedimented meanings of currently existing visions of the future in their almost exclusive focus on hard impacts. This move is akin to what Ihde (2012) calls "to possibillize" the phenomenon, or the "variational method" (23), which we use here not to disclose the one essence of the phenomenon, as classical phenomenology aimed to do, but its potential "multistability"—the fact that technologies acquire different meanings in different contexts. In order to avoid empty speculation and to ensure that possibilities we disclose are really about the phenomenon, we keep our imaginings mundane, geared at a world that must be "intuitable" to our readers (Ihde 2012, 9).

To explore the potential lived experience of automated vehicles, we apply van Manen's (2014) set of existentials or "universal themes of life": lived relation/*relationality*, lived body/*corporeality*, lived space/*spatiality*, lived time/*temporality*, lived things and technology/*materiality*, and lived emotion/*mood* (302–3). We added to these *response-ability*, as a specific form of lived relation, to explore how the lived experience of a world with automated vehicles could impact the foundations of our ethical experience. We understand response-ability in a wide sense, between attentiveness, on the one hand, to feeling ethically responsible or obligated, on the other. The material we present does not always refer to all themes; we used the existentials as a heuristic to brainstorm, together and with others, about possible future lifeworlds.

Van Manen's (2014) exploration of the "vocative"—a text that shows lived experience and makes it felt, instead of telling *about* it—and Ihde's hermeneutic rules for phenomenological observation provided us with some ground rules for the presentation of results from our "possibillizing" exercises in the form of anecdotes. These rules serve to ensure that readers are enabled to "recognize unreflectively" future experiential possibilities presented (van Manen 2014, 241).

- Focus on the lived experience of the phenomenon as it appears. This is done by writing in the present tense, from the perspective of a person experiencing, instead of taking an outsider's perspective;
- focus on describing the experience as such, instead of explaining the phenomenon, drawing the reader into closeness with the phenomenon;
- avoid theoretical terminology and generalizations, while favoring vivid, concrete, experiential detail.

At the same time, the general character of phenomenological observation as oriented to nuances and ambiguities in appearance taught us to stay focused on experience as emerging, not yet constrained by interpretations of its significance.

The aim of the anecdotes we developed is to induce in the reader a sense of wonder at the phenomenon and, where this represents a specific "attitude" to the phenomenon at hand, to invite further exploration of future experiential possibilities. We supplement the anecdotes presented shortly with short reflective commentaries on the significance of—in themselves—relatively mundane experiences recounted.

FOUR PHENOMENOLOGICAL ANECDOTES

(Dis)orientation

> He scans the façades that line the street, and the next, and the next, with their lawns and their flowerboxes. This is their neighborhood. But he has hardly ever passed through here, and certainly not on foot. The car always opts for the ring road. These houses, the families that live here, some of which they must know, he has no sense for how it adds up. Every direction looks the same.
>
> He knows how he got here, closing his laptop with the sales plans he was working on, in response to the beep of the delay alarm. The control screen had predicted a thirteen-minute detour due to the avoidance of unforeseen road obstruction—which usually meant a rally of some kind—to be added to the fourteen-minute time-to-destination left. He had simply jumped out of the car, at the south exit of the ring that circles their neighborhood. It being his daughter's birthday, he had wanted to take things in his own hands. It seemed he was close enough to home to cut through the neighborhood, walking. The car would make it home fine later. Thank heavens it's a clear evening, so he can rely on the last sunlight to keep direction. West, he needs to keep west.
>
> Exhilaration grabs him as he turns the next corner to see their own lawn, their own garage, and, somewhat surprisingly, the car neatly parked in front. Did he take that long? He jogs the last meters home. They must be worried. As he greets their dog, who comes running out as if he were a long lost traveler, it is not just pride he feels. Mostly, it's the silliness of it all. He should walk the dog more often.

At present, people do not usually traverse their own neighborhood by keeping track of the sun. We know our surroundings in an embodied way from navigating them by foot, by bike, by car, involved in shopping, taking children to school, or leisure activities. This is what makes them home. How about the future? Automated vehicles may lead to larger suburban areas (Heinrichs 2016), which would stimulate more motorized travel, not just for commuting, but also for other trips. The anecdote recounts the disorientation in space and time, experienced by the driver of an automated car who leaves his vehicle to walk the last kilometers home by an unfamiliar route. Because he does not know the neighborhood in an embodied way, the space is confusing to him. Being used to time as indicated by technology, he does not have a firm sense of the route home in terms of duration either.

Growing disconnection to spaces traversed, as a possible side effect of automated driving, is not limited to places close to home. With automated cars taking over the navigation, there would be no need for their passengers to build up a stock of visual landmarks, audible clues, or a "feel" for any road traveled. The deskilling this may entail puts travelers at risk when the equipment they rely on stops working—an issue currently surfacing in relation to electronic navigation systems.

Beyond practical issues, the "disorientation" discussed here may also affect the way citizens are "at home" in the world in a more fundamental way. Marc Augé (2009/1995) pointed to the growing number of nonplaces in modern life, like airports, shopping malls, and motorways—places citizens do not inhabit but merely pass through. They figure in these places as anonymous individuals and experience them with a sense of detachment. Will automated vehicles generate their own kind of nonplaces, large areas, far or near, that citizens do not relate to other than as arbitrary transit spaces, their travel time filled with work or activities that have nothing to do with places passed?

Interestingly, when asked about benefits of future automated driving, respondents listed "enjoying the landscape" over other ways of spending traveling time, such as surfing the net, working, or sleeping (Wolf 2016, 118). However, they also fear the pressures of a society oriented to performance in which "lost" time is turned into time for work or dedicated entertainment activities (Fraedrich and Lenz 2016).

What would it mean to be a citizen of a country or province when one no longer connects with its landscape in embodied ways? Will this affect our at-homeness in and our caring for places? And what would happen if we start sending children to school in automated vehicles? In short, we may want to reflect on practices, old and new, that make places meaningful for citizens, both as individuals and as collectives. Designers of autonomous vehicles may want to think about systems that remind users of landmarks or otherwise enable them to develop a connection to space and time traversed.

Public Space as Shared Space

> She loves the city and its freedom. Like this morning, walking to work on an almost empty sidewalk. She always walks, one of the few to do so at commuting time. Against the background of the familiar chorus of whirrs and hums, rising and falling from the strings of vehicles next to her, she has all the time in the world to mull over things she is working on.

> Huddled in her wide coat, hands deep in her pockets, she turns to cross, scanning the passing vehicles. She keeps her eyes down, no need to derange anyone with empty stares. Nor does she have any interest in seeing them, these passengers, bent over their screens for work or some kind of entertainment. The cars are what counts. With their grids flanked by sensors on each side, they appear as endowed with mechanical faces: some dull, some with naïve wide open eyes, while others look positively mean. But the brilliant truth is, there are no differences there, these whirring mechanical creatures are so coolly trustworthy. She doesn't hesitate for a moment and confidently steps forward, as she casually flaps the wing of her coat to make sure she is picked up by the sensor-eyes. As the cars predictably slow down to let her pass, she feels the world is all hers.

The anecdote presents the simple event of a pedestrian crossing a city road. The relationships required to successfully do this are with vehicles, their materiality and sensor systems, not with their passengers. Looking at each other, without clear functionality, is experienced as inappropriate by the protagonist who enjoys the opportunity, provided by automated traffic, to dwell in her private cocoon. The anecdote raises the question what developments like these might mean for the quality of public space, not just as space traversed safely—the main focus of ethics discussing the decision making of automated vehicles—but as "public" or "shared" space.

At present, interaction between participants in traffic is regulated by rules, signs, and electronic traffic control systems. Nevertheless, informal protocols, for instance for allowing a pedestrian to cross or a fellow driver to insert in a lane, play their own role. Within the daily choreography of traffic, we tend to interpret vehicles as extensions of people and respond to their behavior as such. A "pushy" driver may invite a raised fist or angry look, while a courteous one is met with grateful nodding. A slowpoke may be honked at. These interactions make traffic a social space. What happens when a significant part of these interactions is no longer between people but between people and machines?

In *The Fall of Public Man* (1977), Richard Sennett deplores the erosion of public life and the growing self-absorption of citizens, who walk the streets armed by an invisible shield that protects them from the address and looks of others. Though the architecture and organization of urbanized spaces is as much an expression of growing individualism and waning social cohesion as their cause, it is interesting to consider how automated vehicles may impact the social quality of public spaces. Will they, as suggested in the anecdote, contribute to citizens ignoring each other, adding to further anonymization of relationships, possibly even enhancing the much discussed loneliness in modern technologized societies? Or will they bring with them—for example, through extensive car-sharing practices—new kinds of sociality?

Trying to imagine the urban space of the future, it is interesting to consider experiments that have been undertaken since the middle of the last century with traffic space as "shared space." While governments and planners in the first half of the twentieth century sought to enhance safety by segregating different streams of traffic, such as pedestrians, bikers, and automobiles, experiments with streets and squares with little to no regulation and signing suggested safety could also be improved by forcing citizens to observe each others' movements closely and negotiate crossing paths (Hamilton-Baillie 2008). Allegedly, these experiments showed a number of welcome side effects, such as the fostering of civil virtues like courtesy and patience, as well as increased anticipatory and communication skills. What kinds of infrastructure will automated vehicles stimulate? How much space do these structures leave for social interaction?

Another question raised by the anecdote is the ways nonautomated actors may respond to automated vehicles. The protagonist relies on the cars' sensor systems to such an extent that she crosses the road with hardly a look. It is questionable whether one would make such abrupt moves in front of human drivers, if only because it might startle them. How might behavioral changes like these affect traffic safety or flow?

A last reflection pertains to possible stratification of spaces of transit. The introduction of automobiles into the city, replacing horse and buggy, enabled the middle and higher classes to avoid contact with the working poor and immigrants with whom they had shared public transport and the sidewalks in the city (Nikiforuk 2013). Who will occupy the sidewalks in the city of the future? Where and how will citizens meet a diversity of fellow citizens and interact with them? These are questions to consider for policymakers as well as designers of the urban spaces of the future.

Driving/Being Driven

> True, this is safe and comfortable. I trust this car. But sometimes I long back to the wilder days; the confidence that came from bodily intimacy—me, your driver, you, my car. At best, we moved in perfect synchronicity. We complemented each other. I was the one with the eyes, able to direct us through the world. You provided the speed and power, responding to my smallest touch. And in turn, you opened up the world for me in a novel way. You made me experience the road, the speed, the whole exciting jumble of physical forces that propelled us forward. I experienced what you experienced.
>
> But now, with this car, I guess, it takes care of me. Cherishes me like a baby. Or like some old geezer. This car guides me, like a parent. At other times it's my servant. But it certainly stopped being my partner. If I want something from it, I send a message with an order. And it messages and orders me too. We order

each other around. But often I don't really understand this car. Sometimes I have no clue what it tries to tell me. It can act whimsically and refuse my wishes—all for the greater good, of course. My body often feels a bit tense, vigilant—as if preparing for something unexpected to happen. I know it sounds weird, but sometimes I feel alone in this, my, car.

The anecdote aims to capture the change in how we may experience our car, depending on whether we are its driver-partner or its driver-turned-passenger. Driving a vehicle is a favorite example used by phenomenologists to illustrate the so-called embodiment relation we develop with some technologies (for example, Sheller 2004). We relate to the world through these technologies, which enable us to experience that world and to act upon it in particular ways (Verbeek 2005). In some cases our relation with an instrument is so intimate that we no longer experience it as separate from our body; it becomes an integral part thereof. The cane used by a blind person becomes an extension of her fingertips. In the same way, typically, a car becomes an extension of our body. Driving it can become—admittedly not for everyone, and not always—a wordless and intimate dialogue. In an automated car, that intimate dialogue comes to a halt. Such a dialogue is not easy to reinstall. Even if the passenger once in a while still will have to take over as a driver, this will mostly be in the case of emergencies. In those situations, the driver will take back control. (S)he will step in to make corrections. That is: (s)he will be opposing the car, not enjoying the collaboration.

The phenomenological perspective allows us to unpack the well-known but rather broad claim that automated vehicles will take the "fun" out of driving (Fraedrich and Lenz 2016). And in doing so, it may help to problematize all too easy claims that a loss of "fun" should be dismissed as a small price to pay for the economic, environmental, and safety advantages of automatized vehicles. Admittedly, some drivers will not recognize this cozy embodiment relation with their car, yet many others will deplore the loss of this intimacy with their car. Driving a car will stop being a relation that demands the exercise of a skill one can master, that offers a way to express oneself, and that allows for that special type of freedom that consists of just bending the rules a little.

It is important to acknowledge these feelings of loss, as they may initiate unexpected behavior, like a resistance to "comfort and safety" and a refusal to give up the old ways of driving—what technology proponents call "acceptance." Or it could lead to new leisure practices of do-it-yourself driving on remote roads in the countryside or otherwise less developed territories. Policymakers may want to reflect beforehand on the desirability of such new behaviors, while designers could think of ways of safeguarding the special relationship between driver and car that makes a car "my" car.

Unhinged Moral Emotions

> "It was the car, the *car*," the bystanders try to calm her, "there was nothing you could do." But however shocked and shaken they are, she doesn't *feel* like them. She sees how they are taking care of the devastated parents. But she cannot join them. That would be obscene. What words, what gestures could she offer? Apologies would be absurd. But she, of all people, cannot offer consolation or express her sympathies either. She stares at the car—so mute, so unaware, so dumb, so . . . amoral. Only a couple of minutes ago it had cuddled, comforted, and carried her; now it looks cold, distant, treacherous, alien. She feels abandoned by everyone and everything. She just cannot stop thinking of those frightened little eyes and that sickeningly soft "bump" that she alone feels reverberating in her bones.

The anecdote evokes a situation where established moral emotions and attributions of moral responsibility become unhinged as a result of self-driving vehicles.

Our current emotional and practical relationship vis-à-vis moral responsibility is more complex than we think. In the ancient Greek tragedies, we see how a hero like Oedipus was held morally accountable by the Gods for sleeping with his mother and killing his father, even though he did all of this unwittingly. As "moderns" we pride ourselves on only holding actors morally responsible who have knowingly and voluntarily opted for the deeds that caused specific negative outcomes. If there is no willed link between deed and outcome, the language of moral responsibility no longer applies. Or so we tell ourselves.

However, British ethicist Bernard Williams has famously pointed out that "we moderns" actually do hold people accountable for outcomes that are unwilled and that depended on contingent circumstances outside the agent's control. For instance, if someone makes a severe mistake but has the good fortune that no one suffers from it, s/he is condemned much less severely—if at all—than a person who made the same mistake that now, for equally contingent reasons, results in severe harm. He called this phenomenon "moral luck" (Williams 1981). Sometimes we hold people morally responsible, even if it is clear that they couldn't have done anything to avoid a negative outcome. Williams gives the example of a car driver who accidentally runs over a child. Even if we will not hold her legally accountable, on the level of our moral emotions it is clear that her position is quite different from those who were not involved in the accident. This means our actual attitudes toward moral responsibility are considerably less "rational" or "modern" than we like to believe (Williams 2008).

The case presented in the anecdote is more radical. Not only is the lethal accident obviously unwilled, it is even debatable whether there is a causal link between choices made by the "passenger/driver" and the accident.[1] There is no longer a driver whose practical relation to the accident can be explored. That position of the agent is now taken over by the car—even if subsequent evaluations conclude that it cannot be blamed. *But this still doesn't absolve the passenger from experiencing a special moral relation to the accident, and to the victim.* There is the ineffaceable fact that she saw the victim from the perspective of the perpetrator, that she felt its little body under the wheels of the car. There is an indelible intimacy between the victim and the passenger of the car that carries moral weight. An intimacy that in the end rests on another intimacy: that between the car and the passenger. Through its entanglement with the car, the passenger is "tainted" by the death of the child as a result of its collision with the car. She is feeling the utter loneliness of having experienced all this, helplessly, from the perspective of an agent without being one. This loneliness is exacerbated by the fact that she is involved as a companion of a perpetrator, the car, who itself is outside the domain of moral blame and cannot accompany her in her complex moral emotions.

She is also alone because our culture lacks the moral concepts to make sense of these emotions. The standard moral logic of blame and guilt is much too blunt for these new moral experiences. Such concepts utterly fail to make sense of the chaos of feeling simultaneously a perpetrator, a bystander, and a victim. In short, the self-driving car may bring into the world a new moral experience for which the right words, the right emotions, the right practices and rituals still have to be found. She is alone in the sense that our society, as yet, has no rituals available—like a court case, like punishing or begging for and granting of forgiveness—that can start to heal the broken moral world. These new tragedies as yet go without the emotions, words, and rituals to make sense of them.

CONCLUSION

We presented four anecdotes, recounting possible lived experiences of users in a world where automated vehicles have become the norm. These experiences aimed to enlarge the phenomenon "automated vehicles" that in the current ethical literature is limited to an object of safety, security, and comfort—or lacking those qualities. The possibilities we presented are not exhaustive. There are others (for example, the erasure of the current difference in roles between driver and passenger; changing meanings of public transport; the loss of transit time between home and work, when fully automated driving would make it possible to work in the car; the significance of gender differences now often associated with car driving).

The aim of this exercise was rather to sensitize readers to these and other soft impacts of automated vehicles. We presented soft impacts relating to our place in the world ([dis]orientation in space and time), our relation to ourselves and others (public space), our identities in relation to the car as material, mechanical object (driving/being driven), and in relation to the (de)anchoring of our moral emotions (free-floating guilt). In developing these anecdotes we made use of phenomenology as a methodology to help imagine how automated vehicles may affect our identity, our forms of (un)happiness, and our feelings of connectedness with and responsibility toward the environment and to other people. We close with a few remarks on the methodology and on the significance of imagining soft impacts of emerging technologies such as automated vehicles.

Careful attention to the seemingly mundane and openness to nuances and ambiguities in lived experience require expertise. Imagining future lifeworlds and lived experience is supported by familiarity with observing lived experience here and now. Both should follow the common steps of phenomenological observation: the epoche, or stepping back from our ordinary grasp of things, and the hermeneutic rules for phenomenological enquiry: focus on what appears as appearing; describe, do not explain; and remain open to all the experiential content equally, and to the possibilities it entails (Ihde 2012, 18–21). Without these rules, imagining easily slides into an outsider perspective, offering interpretations and sedimented meanings instead of rich lived experience accounts. Exercises in imagining soft impacts, therefore, are ideally supported by phenomenological training in being attentive to lived experience here and now.

To construct an anecdote that presents an experience that did not actually happen, taking place in a lifeworld that has not yet materialized, entails a challenge. How does one step back from experiences one imagined oneself? To avoid that our lived experience accounts became idiosyncratic, we made active use of each other's imaginings and those of various colleagues and acquaintances to test whether experiential possibilities were intuitable and, to an extent, shared. A way to do this more systematically would be to employ focus groups and draw up possible worlds together, after which data from these exercises could be reworked into anecdotes presenting selected experiences.

In the wake of rapid technological development, awareness is growing that technologies do not just come with benefits or risks but have the power to transform our daily lives in unforeseen and wide-ranging ways. Such soft impacts are not easily qualified in terms of an unequivocal good or bad but typically destabilize established interpretative, practical, and normative routines. In doing so they demand answers and adaptations—from ordinary citizens, from businesses and NGOs, from governments. Think of how smartphones have pushed theaters to develop policies reminding users to turn off their phones during performances, railway companies to suggest travelers to adhere to new norms of quiet smartphone use, citizens and businesses to develop new attitudes toward unlimited reachability, and so forth. In all these cases, the soft impacts were dealt with after the fact, and in a happenstance way. Though this is unavoidable to some extent, the exercises in phenomenological imagination presented here aim to train our awareness of soft impacts and their accompanying destabilizations so that we can spot them as soon as they make their appearance rather than glossing them over with our routines until that is no longer possible. In this way, phenomenological training creates space for more timely and reflective responses to our ever-changing techno-social environments.

If engineers and designers are trained in anticipating soft impacts of their designs, by having them imagine how their products may impact the lived experience of users and nonusers alike, this would lead to better design choices and to a degree of moral responsibility that reflects the broad scope of changes technologies effect in the lifeworld. Anecdotes evoking lived experience also form a sobering antidote to the often intoxicating visions of ever-increasing efficiency and comfort that now tend to accompany emerging technologies. Similarly, policy makers need to anticipate technologically induced changes in the lifeworld, as these are often hard to redress once they have manifested themselves. Citizens, once sensitized to possible soft impacts of technologies like automated vehicles, will be empowered to make more informed decisions, not only about their use or nonuse but also about ways of using them and about preferred designs and infrastructures.

In short, imagining soft impacts of new and emerging technologies should be an integral part of discussions on technology development, and a phenomenology of the future is an essential tool to do so. These discussions should not stop with the development phase of technologies but continue after these technologies have entered our lives. Being able to imagine, articulate, and discuss soft impacts should become part of our shared intelligence in dealing with sociotechnological change. After all, few people will be hit by a self-driving vehicle. But everyone's lived experience will be affected by them—for better or for worse.

NOTE

1. Except maybe the causal residue that the car might not have been there at that place and time if she hadn't ordered it. But that is morally no different from someone

who crossed the road before the car two kilometers before, thus similarly codetermining the presence of the car at that time and place.

REFERENCES

Augé, Marc. 2009/1995. *Non-Places. An Introduction to Supermodernity*. London & New York: Verso Books.
Awad, Edmund, Sohan Dsouza, and Richard Kim, et al. 2018. "The Moral Machine Experiment." *Nature* 563: 59–64. https://doi-org.ezproxy.ub.unimaas.nl/10.1038/s41586-018-0637-6
BMVI. 2017a. *Ethik-Kommission Automatisiertes und Vernetztes Fahren*. https://www.bmvi.de/SharedDocs/DE/Publikationen/DG/bericht-der-ethik-kommission.pdf?__blob=publicationFile.
BMVI. 2017b. *Maßnahmenplan der Bundesregierung zum Bericht der Ethik-Kommission Automatisiertes und Vernetztes Fahren (Ethik-Regeln für Fahrcomputer)*. https://www.bmvi.de/SharedDocs/DE/Publikationen/DG/massnahmenplan-zum-bericht-der-ethikkommission-avf.pdf?__blob=publicationFile.
Fraedrich, Eva, and Barbara Lenz. 2016. "Taking a Drive, Hitching a Ride: Autonomous Driving and Car Usage." In *Autonomous Driving. Technical, Legal and Social Aspects*, edited by Markus Maurer, J. Christian Gerdes, Barbara Lenz, and Hermann Winner, 665–85. Berlin: Springer Nature.
Hamilton-Baillie, Ben. 2008. "Shared Space. Reconciling People, Places and Traffic." *Built Environment* 34 (2): 161–81.
Heinrichs, Dirk. 2016. "Autonomous Driving and Urban Land Use." In *Autonomous Driving. Technical, Legal and Social Aspects*, edited by Markus Maurer, J. Christian Gerdes, Barbara Lenz, and Hermann Winner, 213–32. Berlin: Springer Nature.
Ihde, Don. 2012. *Experimental Phenomenology. Multistabilities*. Second edition. Albany: SUNY Press.
Luetge, Christoph. 2017. "The German Ethics Code for Automated and Connected Driving." *Philosophy & Technology* 30 (4): 547–58.
Manen, van, Max. 2014. *Phenomenology of Practice. Meaning-Giving Methods in Phenomenological Research and Writing*. Walnut Creek, CA: Left Coast Press.
Mill, John Stuart. 1859. *On Liberty*. Oxford: Oxford University Press.
Nikiforuk, Andrew. 2013. "The Big Shift Last Time: From Horse Dung to Car Smog. Lessons from an Energy Transition." *The Tyee*, March 6. https://Thetyee.ca/News/2013/03/06/Horse-Dung-Big-Shift.
Santoni De Sio, Filippo. 2016. *Ethics and Self-driving Cars: A White Paper on Responsible Innovation in Automated Driving Systems*. White Paper Commissioned by the Dutch Ministry of Infrastructure and the Environment. TU Delft.
Sennett, Richard. 1977. *The Fall of Public Man*. New York: Alfred A. Knopf.
Sheller, Mimi. 2004. "Automotive Emotions: Feeling the Car." *Theory, Culture & Society* 21: 221–42.
Sokolowski, Robert. 2000. *Introduction to Phenomenology*. Cambridge: Cambridge University Press.

Swierstra, Tsjalling. 2015. "Identifying the Normative Challenges Posed by Technology's 'Soft' Impacts." *Etikk i praksis-Nordic Journal of Applied Ethics* 1: 5–20.

Thompson, Haydn. n.d. "Cyber Physical Systems for Logistics and Transport." In *Ethical Aspects of Cyber-Physical Systems*, edited by Christien Enzing, Chiel Scholten, and Joost Barneveld, 53–63. Brussels: European Parliamentary Research Service. Scientific Foresight Unit STOA.

Verbeek, Peter-Paul. 2005. *What Things Do: Philosophical Reflections on Technology, Agency, and Design*. Pennsylvania State University Press.

Wolf, Ingo. 2016. "The Interaction between Humans and Autonomous Agents." In *Autonomous Driving. Technical, Legal and Social Aspects*, edited by Markus Maurer, J. Christian Gerdes, Barbara Lenz, and Hermann Winner, 103–24. Berlin: Springer Nature.

Williams, Bernard. 1981. *Moral Luck: Philosophical Papers 1973–1980*. Cambridge: Cambridge University Press.

Williams, Bernard. 2008. *Shame and Necessity*. Berkeley and Los Angeles: University of California Press.

Index

accidents. *See* traffic accidents
adaptive regulation, 121–22, 123
air pollution, 102
Alexander, Lamar, 77, 81
algorithms, 4, 19; postphenomenological formula, role of algorithms in, 23, 25; shared ride algorithms, 16, *17,* 21–24, 25
artificial intelligence (AI), 16, 23–24, 35–36, 53–54, 133, 141
Augé, Marc, 157
automated vehicles, 6, 78, 97, 111, 114, 141, 164; in EJ framework, looking backward, 101–3; in EJ framework, looking forward, 103–4, 104–6; ethical literature on, 151–52; future of, 87–88; infrastructure for, 142; phenomenology of, 153–62, 163; responsibilities associated with, 54–55
autonomous vehicles (AVs): autonomous airplanes, 132, 134, 136, 137; autonomous cars, 60–61, 65, 78, 88, 137, 138, 143; autonomous trains, 133–34, 136, 137, 142; banning of non-autonomous vehicles, 5, 60, 61, 65; environmental justice concerns, 6, 97–106; future possibilities for, 111–13; job loss and dislocation issues, 135–36; multiple automated transportation systems, 138–40; potential lived experiences of, 153–56; regulatory guidance for, 113–15, 117–21; soft impacts of, 7, 152, 153, 163–64; trolley scenarios, 3, 11, 19, 131, 142, 151, 152; utopian thinking on, 6, 78, 79–82, 87–90
autonomy, 3, 8, 9, 30, 41, 139; differing levels of, 11, 33, 131; full autonomy, 5, 60, 133
autopilot, 59–60, 62, 68, 134, 137

Belin, Matts-Åke, 3
bicycles, 21, 34–35, 38, 136
Bijker, Weibe, 39
Borenstein, Jason, 7
Brennan, Jason, 68–69, 72
Brey, Philip, 8
Broome, John, 61, 70
Bruntland Report, 25
buses, 22, 104, 142; economic considerations, 135, 143; minibuses, 21, 132, 133; passenger well-being and, 138

Caird, Jeff K., 84
Cane, Peter, 45
Capurro, Raphael, 4

167

carbon emissions, 21, 70, 77, 80–81, 97
cargo ships, 134, 136, 137
Carp, Jeremy, 6–7
causality, 37; backgrounding as a factor, 50–52; causal-responsibility, 44–46; causal singularization, 5, 48–50; causes and causal factors, 46–48; in classification of responsibility, 44–46; foregrounding process and, 52–53
cities, 38, 53, 97, 100, 102, 104, 112; algorithms working in the background of, 21; cities with autonomous vehicles, 33–34, 41, 103; city planning, 6, 25, 98, 139; ethical encounters when crossing city streets, 29; inner-city health concerns, 101; reconfiguring of cities to accommodate AVs, 41, 113; study on eight city transportation modes, 132–34
climate change, 25, 39, 70, 77–78, 80–81
commercial vehicles, 7, 132, 133, 135, 137
composite intentionality, 20–21, 22
COVID-19 pandemic, 2, 11, 122, 133, 134, 141
cybersecurity, 2, 3, 111, 118, 137, 138, 151

Daimler Company, 133
Danaher, John, 63, 74
Deleuze, Gilles, 22, 25
Dewey, John, 79
displacement, 97, 100, 102
DoorDash, 133
driver distraction, 6, 83–89, 90, 112, 136, 137
driverless cars, 11, 39, 61, 110, 113, 114, 120
driverless vehicles, 1, 5, 11, 56
driving: driving automation systems, 33–36, 37, 39, 40; driving impairment and smartphones, 78, 83–84, 85–87, 88, 90; driving risks, 5, 63–65, 68, 70–72, 87; texting while driving, 83–84, 84–85*See also* levels of automation; self-driving cars
drones, 7, 134, 136
Dworkin, Gerald, 45
dynamic driving task (DDT), 34

early warning systems, 8
electric vehicles, 6, 16, 17–19, 21–22, 102
electronic navigation systems, 10, 157
embodiment relations, 17, 160
engines, 4, 16, 17–19, 22, 25, 38
entanglement, 4, 16, 162
environmental justice, 6, 70, 97–106
Epting, Shane, 6
ethics, 32, 139, 158; care, ethics of, 5; causal singularization and, 48–50; of driving risks, 63–65, 70–72; emotionally sentient agents and, 35–36; in environmental justice, 6, 70, 97–106; German code of ethics on AVs, 151, 153; moral theory, 29–30; of multiple AV-types, 134–35; primordial values and, 31, 33; safety as an ethical consideration, 7, 37, 40; of self-driving cars *vs.* regular cars, 65, 66–68, 73; system-level ethics, 60–61, 69, 73, 131–32; of technological change, 7–12; of technology, 8–9, 62–63, 151–52; in Tempe fatal accident case, 34–35; unhinged moral emotions, 161–62

Facebook, 23, 24
The Fall of Public Man (Sennett), 158
FedEx Corporation, 134
field of awareness, 85–87
Figueroa, Robert, 6, 99–100, 101, 103
Firestone Company, 55
Floridi, Luciano, 9
food delivery services, 133
Ford Motor Company, 10–11, 55

Foxx, Anthony, 111
freight, 132, 134, 136, 137, 138
Fritzsche, Albrecht, 4

Gartner hype cycle, 141
General Motors (GM), 133
Geneva Convention on Road Traffic, 118
Georgia Hands-Free Act, 83, 84
Gilmore, Grant, 120
Gogoll, Jan, 61, 69, 72, 74
Goodin, Robert, 45
Google self-driving cars project, 59, 60, 72, 116
greenhouse gases, 25, 77, 81
Grill, Kalle, 64, 71
Guattari, Felix, 22, 25

hacking concerns. *See* cybersecurity
Hansson, Sven Ove, 3, 5, 8, 63–64, 68–70, 72
Hardin, Garrett, 40
Hart, H. L. A., 43–46
Heidegger, Martin, 17, 25
Herkert, Joseph R., 7
hermeneutic relations, 17–18, 20, 155, 163
Hevelke, Alexander, 54
Hicks, Daniel J., 2
horseless carriages, 1, 15, 38, 39
Howard, Mark, 5, 61, 65, 66–68, 72
Hubbard-Mattix, Laci, 105
Hui, Yuk, 2
Human Driving Manifesto, 67, 72
Husak, Douglas, 64
Hydén, Christer, 51
Hyundai Motor Company, 134

Ihde, Don, 16, 17, 19, 24, 154, 155
insurance, 54, 88, 111
intuition, 85, 87, 120

Jackson, Kenneth, 38
justice, 77; automated vehicles and, 101–3, 103–4; distributive justice, 8, 99, 103, 105, 151; environmental justice, 6, 97–99, 99–101; social justice concerns, 7, 138, 139

Kamhof, Ike, 7
Kingdon, John, 115
Kirkman, Robert, 5

Latour, Bruno, 23, 79
law: automated vehicles and the law, 60, 89, 113–14, 123–24; as a barrier to AV transition, 115–16; handheld phone usage as outlawed, 83, 84; technology and the law, 60, 89, 113–14, 123–24; traffic laws, 30, 33, 38
levels of automation, 5, 59, 60, 61–63; conditional driving, 62, 65, 71, 88, 133; full driving automation, 33, 62, 66, 69, 71; high driving level, 11, 62, 71, 133; invisible issues, system-level focus revealing, 131–32; system-level perspective, need for, 142–43
Lin, Patrick, 2–3
lived experience, 9, 10, 152, 164; in phenomenology of ethical action, 30–31; phenomenology of lived experience, 7, 153–56; systems and values discovered in, 41, 162–63
Lundgren, Björn, 3
Lyft, Inc., 11, 113

marginalized communities, 6, 9, 97–106
Marx, Karl, 22
Marx, Leo, 140
McPherson, Tristram, 27
Meadows, Donella, 37–38
mediation, 17, 154
Merleau-Ponty, Maurice, 31, 34
Mill, John Stuart, 47, 152
Miller, Keith W., 7
Mindell, David, 66
Mittelstadt, Brent, 3
Mladenovic, M. N., 53
mobile devices. *See* smartphones

mobility, 1, 4, 109, 112, 132, 142; enhanced mobility, 82, 114–15, 118; mobility networks, improving, 104–5; urban mobility, 102, 106
Moravec's paradox, 12n4
motivational displacement, 33, 35
Müller, Julian, 61, 69, 72
Musk, Elon, 5, 59–60, 61, 65

National Highway Traffic Safety Administration (NHTSA), 2, 112, 113, 119, 122
National Transportation Safety Board, 34–35
Nida-Rümelin, Julian, 54
Nihlén-Fahlquist, Jessica, 64, 71
Nissan Motor Company, 11
Noddings, Nel, 5, 32–33
Noll, Samantha, 105
Nyholm, Sven, 2, 5

obsolescence, 60–61, 62–63, 65, 69, 141
Olsen, Rebecca L., 83

Palm, Elin, 8
paratransit, 132
pedestrians, 4, 5, 19, 20, 36, 158; AVs, pedestrian safety around, 139, 141; ethical encounters between pedestrians and drivers, 29–30; pedestrian crossings, 52–53; recognition between drivers and pedestrians, 32–33, 37; in shared traffic space, 38, 39, 40, 66, 110, 159; in Tempe fatal AV incident, 34–35; traffic accident responsibility and, 47, 50; in trolley problem scenarios, 3
phenomenology, 4, 154; electrical engine and classical postphenomenology, 17–19; ethics of care and, 5; four phenomenological anecdotes, 156–62; lived experience, phenomenology of, 7, 30–31, 153–56; phone usage example, 86–87; postphenomenological analysis, 6; postphenomenological formula, 16, 17–18, 20, 21, 23–24, 25
planes, 132, 134, 135, 136, 137
policy, 101, 111, 116, 119, 122, 124, 152, 159, 160, 164; causality as a factor, 50; in EJ paradigm, 100; future research in AV policy, 104; Georgis Hands-Free Act as good policy, 83; NHTSA policy statement, 113; policy windows, 115; public deliberation and policymaking, 40; regulatory policymaking, 112; shared taxi rides from a policy perspective, 21; soft and hard impacts on policy making, 7; US Department of Transportation, policy vision of, 114

regulation, 7, 16, 79, 87, 113, 119, 125, 143, 152; adaptive regulation, 121–22, 123; business interests, involvement in, 116; environmental regulations, 77, 81; future-facing regulations, 120; jurisdictions, regulation across, 111; law and regulation, 115; loosening grip of, 118; optimal regulation, buying time for, 124; public sector regulation, 110; of ridesharing services, 89; traffic regulation in twentieth century, 159; utopian outlook on, 78; vehicle regulation, 6, 114
relegation, *17*, 21, 24
Reliable Robotics, 134
responsibility: backgrounding as a factor in, 51, 52; blame and task responsibility, 46–47, 50, 52–53, 54–55; Hart, responsibility theory of, 43–46; moral responsibility, 161, 164; response-ability as lived relation, 155
Rice, Stephen, 134
ride sharing. *See* shared rides
risk, 3, 6, 9, 30, 35, 62, 120, 151; in backgrounding examples, 50–51;

bold risk in the digital age, 140–41; conventional driving, riskiness of, 5, 87; equitable system of risk taking, 68–70; offsetting of driving risks, 70–72; at pedestrian traffic crossings, 52–53; philosophy of driving risks, 63–65; safety, taking into consideration, 2, 36–37
Rolls-Royce Motor Cars, 134
Rorty, Richard, 79
Rosenberger, Robert, 6
Roy, Alex, 67

Sabrewing Aircraft, 134
safety, 7, 143; multiple AV-types and traffic safety, 136–37; NHTSA, 2, 112, 113, 119, 122; pro-safety argument for AVs, 2, 4, 36–37, 40, 43, 44, 59; traffic safety, responsibility for, 5, 43–46, 47, 51, 53, 55
Schlosberg, David, 99
Schmidt, Patrick, 6–7
self-driving vehicles. *See* autonomous vehicles
Sennett, Richard, 158
sensors, 16, *17*, 19, 24, 39, 137; LIDAR interference, possibilities of, 139; reliance on as problematic, 158, 159; Tesla sensors and fatal crash, 55
shared rides: autonomous ride-share vehicles, 133, 137; in AV pilot studies, 132; shared-ride algorithms, 16, *17*, 21–24, 25
Shew, Ashley, 81–82
smartphones, 6; driving impairment and, 78, 83–84, 85–87, 88, 90; soft impacts of smartphone use, 152, 153, 164
Smids, Jilles, 2, 64, 71
social experiments, 8
Society for Automotive Engineers (SAE), 5, 34, 61, 88, 133
soft impacts of technology, 7, 152–53, 163–64

Sparrow, Robert, 5, 54, 61, 62, 65, 66, 67, 72
spectatorial utopianism, 6, 78, 81, 82
Stirk, Peter, 141
Stivers, Richard, 141
Strayer, David L., 85
subtechnology families: electric engines, 16, 17–19, 24; self-driving mechanisms, 16, 19–21; shared-ride algorithms, 16, *17*, 21–24, 25
Swierstra, Tsjalling, 7
systems, 9, 22, 39, 55, 84, 87, 89, 137, 142, 157; adaptive systems, 122–23; automation systems, 54, 61–62, 63, 65; communication systems, 2, 3, 54, 55, 56; current systems, 98, 101, 104; driver assistance systems, 59, 67; driving automation systems, 36–37, 39, 40; mobility systems, 1, 4, 102; multiple automated transportation systems, 138–40; public transport systems, 100, 138; of risk taking, 69, 72; sensor systems, 158, 159; sociotechnical systems, 7, 131, 143; technological systems, 3–4, 37–38; transportation systems, 1, 97–98, 100, 102–3, 105, 132, 140, 142

taxi services: flying taxis, 134–35, 136, 137, 138, 139, 141, 143; as mobility-on-demand, 132; shared taxi rides, 21–22, 24, 25; taxi drivers and potential job loss, 112–13, 135
technological progress, 6, 78, 81, 82, 125
technology: future technology, concerns over, 109, 111–13; pacing problem with new technology, 110, 115, 118; philosophy of technology, 16, 79, 80, 86; planned adaptive approach to adopting, 121–22; spectatorial utopian thought on, 6, 78, 81–82, 87, 89–90; technological accidents, 141; technological change, 7–11, 16, 79, 110–11, 115, 117–21,

164; technological fix for social problems, 140; technological frames, 36–39; technological intentionality, 4, *17,* 19–21, 22, 23, 24–25; technological systems, 3–4, 9–10, 37–38; technology ethics, 8–9, 62–63, 151–52

Tesla Model S cars, 55, 59–60, 62, 68

texting while driving, 83–84, 84–85

traffic accidents: causal singularization as a factor, 48–50; driver distraction, 6, 83–90, 112, 136, 137; driving risks and, 64, 68; highly visible accidents, 123; large-scale accidents, 136; self-driving accidents with fatalities, 133; Tempe, fatal accident in, 34–35; unhinged moral emotions caused by, 161–62

traffic fatalities, 2, 34–35, 52, 133

trains, 133–34, 136, 137, 142

transportation, 37, 40, 137; employment in transportation sector, 135; multiple automated transportation systems, 138–40; National Transportation Safety Board, 34–35; public transport systems, 100, 138, 142; shared rides and, 21–24; study on eight city transportation modes, 132–34; transportation systems, 1, 97–98, 100, 102–3, 105, 132, 140, 142; US Department of Transportation, policy vision of, 114; Vision Zero on transportation utilization, 53

trolley problem scenarios, 3, 11, 19, 131, 142, 151, 152

trucks: in accidents, 47, 55; autonomous commercial trucks, 133, 137, 138–39; economic considerations, 88, 118, 135; long-haul trucking, 64, 113–14, 132, 136

Trump, Donald, 77

Uber, 113, 134

Urmson, Chris, 59–60, 61, 62, 65, 72

Van de Poel, Ibo, 8

Van Manen, Max, 154, 155

Verbeek, Peter-Paul, 4, 16, 19–20

vernacular value visioning (VVV), 9–10

Vision Zero project, 5, 53, 55

Volante, Alisha, 100

Volvo Cars, 3, 68, 116

Walmart, Inc., 133

Waymo Company, 133

Weinberg, A. M., 140

Wellner, Galit, 4

Weston, Anthony, 33

"When Humans Are Like Drunk Robots" (Sparrow/Howard), 65

Williams, Bernard, 161

Winner, Langdon, 140

Winter, Scott, 134

Wiseman, Yair, 141

Wittgenstein, Ludwig, 50

Yang, Andrew, 141

Zuboff, Shoshana, 24, 138

About the Contributors

Jason Borenstein, PhD, is the director of graduate research ethics programs at the Georgia Institute of Technology. His appointment is divided between the School of Public Policy and the Office of Graduate Studies. He has directed the Institute's Responsible Conduct of Research (RCR) Program since 2006. Dr. Borenstein is also chair of the Association for Practical and Professional Ethics Research Integrity Scholars and Educators (APPE RISE) Consortium and a member of the IEEE SSIT Technical Committee on Ethics/Human Values. He is an international editorial advisory board member of the Springer journal *Science and Engineering Ethics*, a founding editorial board member of the Springer journal *AI and Ethics*, coeditor of the *Stanford Encyclopedia of Philosophy*'s Ethics and Information Technology section, and an editorial board member of the journal *Accountability in Research*. His teaching and research interests include robot and AI ethics, engineering ethics, research ethics/RCR, and bioethics.

Jeremy Carp is an associate at Perkins Coie LLP. He previously clerked for the Honorable Julia S. Gibbons of the US Court of Appeals for the Sixth Circuit and the Honorable Michael J. McShane of the US District Court for the District of Oregon. Jeremy received his law degree magna cum laude from the University of Pennsylvania and his bachelor's degree summa cum laude from Macalester College. He lives in Portland, Oregon.

Shane Epting is assistant professor of philosophy at the Missouri University of Science and Technology. He teaches courses such as "Transportation Justice," "Philosophy of the City," and "Creating Future Cities." Epting cofounded the Philosophy of the City Research Group. Rowman & Littlefield published his latest work, *The Morality of Urban Mobility: Technology and Philosophy of the City*.

Sven Ove Hansson, professor of philosophy at the Royal Institute of Technology, Stockholm, is the editor-in-chief of *Theoria*, a member of the Royal Swedish Academy of Engineering Sciences, and past president of the Society for Philosophy and Technology. He has published seventeen books and about 390 peer-reviewed papers on moral and political philosophy, the philosophy of science and technology, logic, decision theory, and the philosophy of risk.

Joseph Herkert, DSc, is associate professor emeritus of science, technology, and society at North Carolina State University. He has been teaching engineering ethics and STS for thirty-five years. He is editor of two books and has published in engineering, law, social science, and applied ethics journals and edited volumes. Herkert previously served as editor of *IEEE Technology and Society Magazine* and as associate editor of *Engineering Studies*. He is a fellow of the American Society for Engineering Education, a fellow of the American Association for the Advancement of Science, and a life senior member of IEEE. Recent work includes ethics of autonomous vehicles, lessons learned from the Boeing 737 MAX crashes, and responsible innovation in biotechnology.

Ike Kamphof is researcher and teacher in the Department of Philosophy of the Faculty of Arts and Social Sciences at Maastricht University. Her work focuses on ethics in technologically mediated networks of care. She published on health care for vulnerable elderly people and nature conservation. Ike's work combines (post)phenomenology with ethnography and virtual ethnography of user practices. Recently, she was project leader of the funded research project Make-Believe Matters: The Moral Role Things Play in Dementia Care.

Robert Kirkman is associate professor and director of graduate studies for advising in the School of Public Policy at Georgia Tech. His research centers on practical ethics and related issues in phenomenology, cognitive psychology, and technology studies. Past work has focused on environmental ethics and policy, especially the ethics of the built environment as it plays out in metropolitan regions of the United States, and on ethics education and assessment. His current project explores the affinity between the experience of making music with and for others and the prereflective meaning of ethical action. He is the author of *The Ethics of Metropolitan Growth: The Future of Our Built Environment* (Continuum, 2010) and numerous journal articles, including publications in *Environmental Ethics, Environmental Values, Advances in Engineering Education, Teaching Ethics*, and *Science, Technology & Human Values*.

Diane Michelfelder is professor of philosophy at Macalester College. Her primary area of research is the ethics of emerging networked technologies. She has served the Society for Philosophy of Technology in multiple roles, including as president and as co-editor-in-chief of the society's journal, *Techné: Research in Philosophy and Technology*. She was also actively involved in the formation and development of the Forum on Philosophy, Engineering, and Technology (fPET) and is a member of its steering committee. Her work has been published in this journal as well as in *Science and Engineering Ethics, AI & Society, Philosophy and Technology, Engineering Studies*, and *Ethics and Information Technology*, among others. Her most recent book is the *Routledge Handbook of the Philosophy of Engineering*, edited with Neelke Doorn (Routledge, 2020).

Keith W. Miller is a professor at the University of Missouri–St. Louis, where he is a member of two faculties: the Department of Computer Science (PhD in computer science in 1983) and the College of Education (BS in education in 1973). In 2013, Dr. Miller became the Orthwein Endowed Professor at the University of Missouri–St. Louis. As part of that endowment, he has the Saint Louis Science Center as his community partner. He also collaborates with Girls' Inc. of St. Louis and with the nonprofit Launchcode. Prof. Miller teaches software engineering and computer ethics and mentors EdD and PhD students in the College of Education. Dr. Miller has published research on computer ethics since 1988. For more details about Prof. Miller, please see http://learnserver.net/faculty/keithmiller. For an analysis of citations to his research, please see https://scholar.google.com/citations?user=egfNDGwAAAAJ&hl=en.

Sven Nyholm is an associate professor at the Ethics Institute of Utrecht University in the Netherlands. He is also a member of the ethics advisory board of the Human Brain Project and an associate editor of the journal *Science and Engineering Ethics*. Nyholm's main research areas are ethical theory and applied ethics, especially the ethics of technology. Lately, much of his research has been about philosophical issues related to robots and artificial intelligence. Nyholm's most recent book is *Humans and Robots: Ethics, Agency, and Anthropomorphism* (Rowman & Littlefield International, 2020).

Robert Rosenberger is associate professor of philosophy in the School of Public Policy at the Georgia Institute of Technology, and he is president-elect of the international Society for Philosophy and Technology. His research includes the development of the postphenomenological philosophical perspective, as well as long-term case studies into topics such as scientific imaging, smartphone-induced driving impairment, and hostile design and architecture. He is a cofounder and the editor-in-chief of the Lexington Books series Postphenomenology and the Philosophy of Technology. Rosenberger is the author of *Callous Objects: Designs Against the Homeless* and editor or coeditor of *Postphenomenological Investigations, Philosophy of Science: 5 Questions* and *Postphenomenology and Imaging: How to Read Technology*.

Patrick Schmidt is professor of political science and codirector of legal studies at Macalester College. His teaching and research ranges within American public law and regulation, with particular interest in the legal profession and constitutional law. He completed a PhD in political science at the Johns Hopkins University and was the John Adams Research Fellow at the Centre for Socio-Legal Studies, Oxford. He is the author or editor of four books, including *Lawyers and Regulation: The Politics of the Administrative Process* (Cambridge).

Tsjalling Swierstra is a philosopher of technology at Maastricht University, the Netherlands. He is initiating cofounder of the *Journal for Responsible Innovation* and has published widely on the ethics of new and emerging science and technology (NEST-ethics), on the "soft impacts" of technology, and on technomoral change, that is, the mutual shaping of science, technology, and morals.

Galit Wellner, PhD, is an adjunct professor at Tel Aviv University and at Bezalel Academy of Art and Design. Galit studies digital technologies and their interrelations with humans. She is an active member of the Postphenomenology Community that studies the philosophy of technology. She has published several peer-reviewed articles and book chapters. Her book *A Postphenomenological Inquiry of Cellphones: Genealogies, Meanings and Becoming* was published in 2015 (Lexington Books). She translated Don Ihde's book *Postphenomenology and Technoscience* to Hebrew (Resling, 2016). She also coedited *Postphenomenology and Media: Essays on Human–Media–World Relations* (Lexington Books, 2017). Recently she has published several articles and book chapters on gender bias in AI and algorithmic imagination. She is currently working on two books on attention and imagination.

www.ingramcontent.com/pod-product-compliance
Lightning Source LLC
Chambersburg PA
CBHW021850300426
44115CB00005B/98